Electric Power Systems Manual

Other Electrical Power Engineering Books of Interest

FINK AND BEATY • *Standard Handbook for Electrical Engineers*
KURTZ AND SHOEMAKER • *The Lineman's and Cableman's Handbook*
KUSKO • *Emergency/Standby Power Systems*
LINDEN • *Handbook of Batteries and Fuel Cells*
LUNDQUIST • *On-Line Electrical Troubleshooting*
SEIDMAN, MAHROUS, AND HICKS • *Handbook of Electric Power Calculations*
SMEATON • *Switchgear and Control Handbook*

Electric Power Systems Manual

Geradino A. Pete, P.E.
Vector Engineering Services

McGraw-Hill, Inc.
New York   St. Louis   San Francisco   Auckland   Bogotá
Caracas   Lisbon   London   Madrid   Mexico   Milan
Montreal   New Delhi   Paris   San Juan   São Paulo
Singapore   Sydney   Tokyo   Toronto

Library of Congress Cataloging-in-Publication Data

Pete, Geradino A.
 Electric power systems manual / Geradino A. Pete.
     p.   cm.
 Includes index.
 ISBN 0-07-049530-0
  1. Electric power systems—Handbooks, manuals, etc. I. Title.
 TK1001.P47 1992
 621.319′1—dc20   91-48339
                                                                  CIP

Copyright © 1992 by McGraw-Hill, Inc. All rights reserved. Printed in the United States of America. Except as permitted under the United States Copyright Act of 1976, no part of this publication may be reproduced or distributed in any form or by any means, or stored in a data base or retrieval system, without the prior written permission of the publisher.

1 2 3 4 5 6 7 8 9 0   DOC/DOC   9 8 7 6 5 4 3 2

ISBN   0-07-049530-0

The sponsoring editor for this book was Harold B. Crawford, the editing supervisor was Kimberly A. Goff, and the production supervisor was Donald F. Schmidt.

Printed and bound by R. R. Donnelley & Sons Company.

Information contained in this work has been obtained by McGraw-Hill, Inc., from sources believed to be reliable. However, neither McGraw-Hill nor its authors guarantees the accuracy or completeness of any information published herein and neither McGraw-Hill nor its authors shall be responsible for any errors, omissions, or damages arising out of use of this information. This work is published with the understanding that McGraw-Hill and its authors are supplying information but are not attempting to render engineering or other professional services. If such services are required, the assistance of an appropriate professional should be sought.

To Lois, Doug, Danielle, Kailee, and all of my family

Contents

Preface xi
Acknowledgments xiii

Chapter 1. Frequency, Voltage, and Current 1
  1.1 Power System Frequency 1
  1.2 Measurement of Power System Frequency 6
  1.3 Sinusoidal Voltage and Current 9
  1.4 Value of Sinusoidal Voltage and Current 10
  1.5 Measurement of RMS Quantities 14
  1.6 Digital Computation of RMS Quantities 15
  1.7 Average Value of Sinusoidal Voltage and Current 17
  1.8 Voltage and Current Magnitudes 19
  1.9 Voltage and Current Phasors 20
  1.10 Angular Relationships 23
    1.10.1 Electrical Length 23
    1.10.2 Phase Angles 25
    1.10.3 Voltage Angles 25
    1.10.4 Current Angles 26
  1.11 Three-Phase Voltage and Current 27
  1.12 Power System Harmonics 30

Chapter 2. Power 33
  2.1 Instantaneous Power 33
  2.2 Real Power (P) 35
  2.3 Reactive Power (Q) 37
  2.4 Total Power 38
  2.5 Complex Power (S) 39
  2.6 Power Factor 40
    2.6.1 Leading Power Factor 40
    2.6.2 Lagging Power Factor 41
  2.7 Energy 42
  2.8 Measurement of Real and Reactive Power 43

Chapter 3. Circuit Elements　　45

- 3.1 Resistance (R)　　45
  - 3.1.1 Voltage and Current Relationships　　46
  - 3.1.2 Power in Resistance　　47
- 3.2 Inductance (L)　　49
  - 3.2.1 Voltage and Current Relationships for Inductance　　50
  - 3.2.2 Inductive Reactance　　51
  - 3.2.3 Power in Inductance　　53
  - 3.2.4 Energy in Inductance　　54
- 3.3 Capacitance (C)　　57
  - 3.3.1 Voltage and Current Relationships　　58
  - 3.3.2 Capacitive Reactance　　60
  - 3.3.3 Power in Capacitance　　61
  - 3.3.4 Energy in Capacitance　　63
- 3.4 Impedance and Admittance　　65

Chapter 4. Transmission Lines　　67

- 4.1 Voltage and Current as Functions of Distance　　68
  - 4.1.1 Development of the Voltage Equation　　68
  - 4.1.2 Development of the Current Equation　　70
  - 4.1.3 Solution of the Transmission Line Equations　　71
- 4.2 Voltage and Current as Functions of Distance and Time　　75
  - 4.2.1 Development of the Voltage Equation　　76
  - 4.2.2 Development of the Current Equation　　76
  - 4.2.3 Substitution of Derivatives　　77
  - 4.2.4 Solution of the Voltage and Current Equations as Functions of Distance and Time　　78
- 4.3 Terminal Conditions for the Transmission Line Equations　　81
  - 4.3.1 Receiving-End Voltage and Current　　82
  - 4.3.2 Sending-End Voltage and Current　　83
  - 4.3.3 Transmission Line Constants (A, B, C, and D)　　84
- 4.4 Equivalent Transmission Line Models　　84
  - 4.4.1 Equivalent Pi Model and T Model　　85
  - 4.4.2 Approximate Pi Model and T Model　　87
- 4.5 Characteristic Impedance　　88
  - 4.5.1 The Lossy Case　　89
  - 4.5.2 The Lossless Case　　91
- 4.6 Propagation Constant　　91
  - 4.6.1 Lossy Case　　92
  - 4.6.2 The Lossless Case　　94
  - 4.6.3 The Distortionless Case　　95
- 4.7 A Practical Example　　95

Chapter 5. Power System Loads　　105

- 5.1 Load Bus Specifications　　105
- 5.2 Transmission Line Specifications　　108
- 5.3 Source Bus Specifications　　109
- 5.4 Power Factor Correction　　110
- 5.5 Transmission System Efficiency　　112

5.6	Transmission Line Voltage Regulation	114
5.7	Load Factor	114

Chapter 6. Power Circle Diagrams 117

6.1	Derivation of the Power Circle Diagram Equations	118
6.2	A Practical Example of Real and Reactive Power Flow	120
6.3	The Relationship between Reactive Power Flow and Voltage Magnitude	124
6.4	The Relationship between Real Power Flow and Voltage Angle	126
6.5	Surge Impedance Loading	127
6.6	The Power Angle Diagram	129

Chapter 7. Symmetrical Components 131

7.1	Introduction	131
7.2	Fault Point	132
7.3	Positive Sequence Components	133
	7.3.1 Positive Sequence Current	133
	7.3.2 Positive Sequence Voltage	133
	7.3.3 Positive Sequence Impedance	134
7.4	Negative Sequence Components	134
	7.4.1 Negative Sequence Current	134
	7.4.2 Negative Sequence Voltages	135
	7.4.3 Negative Sequence Impedance	135
7.5	Zero Sequence Components	136
	7.5.1 Zero Sequence Current	136
	7.5.2 Zero Sequence Voltage	136
	7.5.3 Zero Sequence Impedance	137
7.6	Phase Quantities Versus Sequence Quantities	137
7.7	Line-to-Line Fault	139
7.8	Line-to-Ground Fault	146
7.9	Double-Line-to-Ground Fault	153
7.10	Applications to System Protection	159

Chapter 8. Transient Analysis 161

8.1	Undamped Case	163
8.2	Critically Damped Case	166
8.3	Underdamped Case	169
8.4	Overdamped Case	172
8.5	Energy Considerations	175
8.6	A Practical Example	176

Chapter 9. Symmetrical Versus Asymmetrical Current 181

9.1	Initial Conditions	181
9.2	Derivation of the Total Current Equation	182
	9.2.1 Complementary Solution (i_C)	184
	9.2.2 Particular Solution (i_P)	185
	9.2.3 Complete Solution ($i = i_C + i_P$)	186
9.3	Fault Analysis	187

Chapter 10. Transformers — 191
- 10.1 Basic Transformer Theory — 191
- 10.2 Transformer Flux Relationships — 192
- 10.3 The Ideal Transformer — 194
- 10.4 Transformer Action — 195
- 10.5 Turns Ratio — 197
- 10.6 Coefficient of Coupling — 197
- 10.7 Primary and Secondary Quantities — 198
- 10.8 The Practical Transformer — 200
- 10.9 Transformer Impedances — 200
    - 10.9.1 Winding Resistance — 200
    - 10.9.2 Winding Reactance — 200
    - 10.9.3 Core Resistance — 200
    - 10.9.4 Magnetizing Reactance — 201
- 10.10 Reflected Primary and Secondary Quantities — 201
    - 10.10.1 Secondary Quantities Reflected to the Primary — 201
    - 10.10.2 Primary Quantities Reflected to the Secondary — 202
- 10.11 Transformer Excitation — 204
- 10.12 Saturation — 206
- 10.13 Delta-Wye Transformations — 207
    - 10.13.1 Current Relationships — 207
    - 10.13.2 Voltage Relationships — 210
- 10.14 Current Transformers — 212
    - 10.14.1 Current Transformer Model — 213
    - 10.14.2 Burden — 213
    - 10.14.3 Performance Under Normal Conditions — 213
    - 10.14.4 Performance Under Fault Conditions — 214

Appendix. Solution Methods for Transmission Line Equations — 217
- A.1 Method of Laplace Transform — 217
    - A.1.1 Solution of the Voltage Equation — 217
    - A.1.2 Solution of the Current Equation — 219
- A.2 Method of Series Expansion — 221
    - A.2.1 Solution of the Voltage Equation — 221
    - A.2.2 Solution of the Current Equation — 224

Index 227

Preface

The *Electric Power Systems Manual* is intended to provide electrical engineers in the electric power industry with an organized approach to power systems analysis. The content and organization of this book are based on the courses which I have conducted for the engineers and technicians of the electric utility industry. The book was originally conceived as a personal notebook which I compiled to record methods of analysis for topics associated with electric power engineering.

The Fourier series, for example, is used to show why a delta winding may be used to block third harmonics currents. Symmetrical components are used to show why the delta winding represents an open circuit to zero sequence components associated with ground faults. Power circle diagrams are used to illustrate the relationship between real power flow and voltage angle versus the relationship between reactive power flow and voltage magnitude. Classical methods of differential equations are used to develop the steady-state and transient components of asymmetrical fault current. The electrical engineer who enters the field of power engineering without the benefit of an academic background in power systems analysis will be pleased to note that he or she already possesses the analytical skills that are needed to understand these concepts.

The first three chapters are designed to provide the reader with a review of the fundamental concepts of electrical engineering in context with power system applications. These concepts include frequency, frequency deviation, and frequency control as they relate to the electric power system; the use of phasor representation as the language of the electric power engineer; the relationships between voltage, current, and impedance associated with power system generation, transmission, and loads; and the development of real and reactive power principles which are so important in understanding power system operation.

Chapter 4, "Transmission Lines," includes a rigorous treatment of the development, solution, and application of transmission line theory. This analysis will prepare the electric power engineer for the application of "long line" theory to operating transmission lines.

Chapter 5, "Power Circle Diagrams," provides an in-depth view of a most useful analytical tool for understanding the relationships between the magnitudes and angles of bus voltages to real and reactive power flow. This tool is particularly useful to the system operations engineer who is responsible for making daily decisions regarding generator dispatch, switching of reactive power devices, and loading of the transmission system and static stability.

Chapter 6, "Power System Loads," examines the practical considerations associated with the delivery of electrical service to a large customer. The improvements in operating efficiency and voltage regulation that can be achieved via power factor correction is analytically demonstrated.

Chapter 7, "Symmetrical Components," includes actual screen captures of pre-fault and post-fault voltages and currents which were obtained from the electric power system with the use of a digital fault recorder. The theoretical development of symmetrical components is related to the actual quantities obtained during the fault.

Chapter 8, "Transient Analysis," relates the various categories of undamped and damped voltages and currents to the natural response of the electric power system. A practical example of the natural response of a 765kV transmission line following line clearing is examined.

Chapter 9, "Symmetrical Versus Asymmetrical Current," examines a topic which is essential to the establishment of momentary ratings for power system devices. The basic concepts of circuit analysis are extended to a practical example using a faulted 345kV transmission line with a graphic illustration of the steady-state, transient, and total fault current and the time constant associated with the unidirectional component.

Chapter 10, "Transformers," includes a discussion of the benefits in using the ideal transformer as the basis of the development of practical models for instrument and power transformers. The importance of understanding such concepts as leakage reactance, ratio and phase angle correction, real power losses, and saturation of power and instrument transformers is emphasized.

I would like to extend best wishes to the electric power engineer in the pursuit of engineering excellence. I hope that the material presented in the *Electric Power Systems Manual* will help you to achieve that goal.

Geradino A. Pete, P.E.

Acknowledgments

I would like to acknowledge the efforts of the following people who have contributed directly or indirectly to the development of the Electric Power Systems Manual.

The members of the System Protection Technician Training Task Force of the New York Power Pool for the guidance, patience, and understanding which they extended to me during the development of this manuscript. They include Cliff Tienken of the Central Hudson Gas & Electric Company; Al Seul of Consolidated Edison; Andy Maksimchak of the Long Island Lighting Company; Richard Alfano and Steve Smith of the New York Power Authority; Jim Ingleson of the New York Power Pool; Bill Kahrl of the New York State Electric & Gas Corporation; Bob O'Brien of the Niagara Mohawk Power Corporation; Frank Kaczmark, Jr., of Orange & Rockland Utilities; and Ralph J. Schwartz of the Rochester Gas & Electric Corporation

The industry sponsors who provided encouragement to me regarding the development of this book. They include William J. Balet of the New York Power Pool; Melvin I. Olken of the Institute of Electrical and Electronic Engineers; and Mike Henderson, Lew Burnett, Lenny Panzica, and John Phillips of the New York Power Authority

Walid Hubbi, Ph.D., Associate Professor, Electrical and Computer Engineering Department of the New Jersey Institute of Technology whose guidance and suggestions were invaluable to me in the development of this book.

Geradino A. Pete, P.E.

Electric Power Systems Manual

CHAPTER 1

FREQUENCY, VOLTAGE, AND CURRENT

1.1 POWER SYSTEM FREQUENCY

The *frequency f* of a sinusoidal function of time represents the number of complete cycles traversed by the function in a period of 1 second (s). The frequency is measured in units of cycles per second (cps) or hertz[1] (Hz). The reciprocal of the frequency is the *period* τ which represents the amount of time in seconds that is required for the function to complete one full cycle. The voltages and currents which are produced by the generators of the electric power system are sinusoidal in nature.

These voltages and currents give rise to electric power which is also sinusoidal in nature. The scheduled frequency for the voltage and current on the electric power system in most of North and South America is

$$f = 60 \text{ Hz} \qquad (1\text{-}1)$$

The period of a 60-Hz sinusoidal voltage or current is

$$\tau = \frac{1}{60} = 0.0167 \text{ s} \qquad (1\text{-}2)$$

Time in the electric power industry may be measured in terms of cycles. A cycle is a unit of time which represents the period of a 60-Hz wave. An event which persists for 1 s, for example, is said to last for 60 cycles. High-speed devices can respond to events such as faults on the electric power system within a few cycles.

---

[1]The unit hertz (Hz) as a measure of frequency is named in honor of the German physicist Heinrich Rudolph Hertz (1857-1894) who demonstrated that electromagnetic radiation results from the oscillations of electricity in a conductor.

CHAPTER 1

The *radian frequency* ω of a sinusoidal function is the number of radians traversed by the function in a period of 1 s.

A *radian* is a unit of angle measurement which describes an arc on the circumference of a circle that is equal in length to the radius. Since the radius traverses 360° in a complete cycle, the number of degrees described by an arc of 1 radian (rad) is

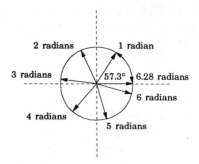

Figure 1-1 Radians.

$$\theta = \frac{360°}{2\pi \text{ rad}}$$
$$= 57.3° \text{ per rad}$$

The Greek letter π represents a constant in mathematics which defines the relationship between the diameter D of a circle and the circumference C.

$$C = \pi D \tag{1-3}$$

Since the radius of a circle is half of the diameter, the relationship between the radius and the circumference is

$$C = 2\pi R \tag{1-4}$$

This result stems from the integration of the product of the arc length per radian and the number of radians in a complete cycle.

$$C = \int_0^{2\pi} R\, d\theta = R\theta \Big|_0^{2\pi} = R(2\pi - 0) = 2\pi R \tag{1-5}$$

Table 1-1 is an example of the relationship between degrees and radians for representative angles.

FREQUENCY, VOLTAGE, AND CURRENT

Table 1-1 Cycles, Degrees, and Radians

CYCLES	DEGREES	RADIANS
¼ -cycle	90	$\pi/2$
½ -cycle	180	π
¾ -cycle	270	$3\pi/2$
1 cycle	360	2π

The radian frequency ω is related to the scheduled power system frequency of 60 cps or 60 Hz as follows:

$$\omega = \frac{2\pi \text{ rad}}{\text{cycle}} \cdot \frac{60 \text{ cycles}}{\text{s}} = 377 \frac{\text{rad}}{\text{s}} \qquad (1\text{-}6)$$

A balance must be maintained between the generation on the electric power system and the sum of the loads and losses in order to maintain the scheduled power system frequency of 60 Hz.

$$P_{gen} = P_{load} + P_{losses} \qquad (1\text{-}7)$$

Automatic generation control (AGC) schemes are employed to maintain the desired levels of generation and frequency. A deficiency of generation will result in an actual power system frequency which is less than the scheduled value; conversely, an excess of generation will result in an actual power system frequency which is greater than the scheduled value. The *frequency bias mode* of AGC monitors the deviation from the scheduled system frequency and adjusts the available generation to cancel the measured deviation.

The frequency bias is generally expressed as the amount of real power in megawatts (MW) that is necessary to produce a change in the power system frequency of one-tenth of a hertz. Assume that the scheduled power system frequency is 60 Hz and the actual frequency has declined to 59.98 Hz.

CHAPTER 1

The frequency deviation is

$$\Delta f = f_{actual} - f_{scheduled}$$
$$= 59.98 - 60.00 \quad (1\text{-}8)$$
$$= -0.02 \text{ Hz}$$

Assume that the frequency bias for this control area is 20 megawatts (MW) per tenth of a hertz. The amount of additional generation that is necessary to restore the power system frequency to the scheduled value is

$$\Delta P = \frac{\Delta P}{\Delta f} \cdot \Delta f$$
$$= \frac{20 \text{ MW}}{0.1 \text{ Hz}} \cdot 0.02 \text{ Hz} \quad (1\text{-}9)$$
$$= 4 \text{ MW}$$

Another aspect of system operation that involves the maintenance of the scheduled system frequency is *time error correction*. The time that is recorded by clocks which are synchronized to the frequency of the electric power system is compared to the time that is recorded by the time standards of the North American Electric Reliability Council. Any frequency deviation will cause a difference in these recorded times that must be corrected by an adjustment in the level of generation.

Consider that the frequency deviation of -0.02 Hz cited above persists for a period of 1 hour (h). The cumulative discrepancy in the number of cycles is

$$\frac{-0.02 \text{ cycles}}{s} \cdot \frac{3600 \text{s}}{h} = -72 \frac{\text{cycles}}{h} \quad (1\text{-}10)$$

The integrated time error over this period is

$$\frac{-72 \text{ cycles}}{h} \cdot \frac{1 \text{ s}}{60 \text{ cycles}} = -1.2 \frac{s}{h} \quad (1\text{-}11)$$

FREQUENCY, VOLTAGE, AND CURRENT

Table 1-2 NERC Time Correction Limits *(Courtesy of the North American Electric Reliability Council)*

TIME CORRECTION	TIME OF INITIATION	INITIATION TIME ERROR (s)			TERMINATION TIME ERROR (s)		
		EAST	WEST	ERCOT	EAST	WEST	ERCOT
Slow	0000-0400	-4	-2	-3	0	±0.5	±0.5
	0400-2000	-8			-4		
	2000-2400	-4			0		
Fast	0000-0400	+8	+2	+3	+4	±0.5	±0.5
	0400-1200	+4			0		
	1200-1700	+8			+4		
	1700-2000	+4			0		
	2000-2400	+8			+4		

The limits which are established by the North American Electric Reliability Council (NERC) for time correction within the East and West systems and the Electric Reliability Council of Texas (ERCOT) are listed in Table 1-2. The interconnection monitors of the NERC are required to inform the Regional Monitors when the integrated time error reaches a predetermined number of seconds in order that corrective measures may be initiated.

It has been noted that the negative time error that occurs during peak-load periods due to underfrequency conditions may be offset by the positive time error that occurs during light-load periods due to overfrequency conditions. This minimizes the need for intervention by the member systems. The negative time error that results from an underfrequency condition is referred to as a *slow* condition; conversely, the positive time error that results from an overfrequency condition is referred to as a fast condition.

CHAPTER 1

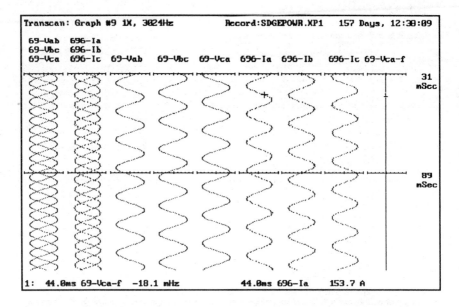

Figure 1-2 Digital fault recorder scan. (*Courtesy of Mehta Tech, Inc.*)

1.2 MEASUREMENT OF POWER SYSTEM FREQUENCY

The measurement of deviation from the scheduled power system frequency of 60 Hz for an operating 69-kilovolt (kV) subtransmission line is illustrated in Fig. 1-2. This information was captured with a digital fault recording system. We will conduct a thorough examination of the output from this device since it will be used to provide actual data recorded from the electric power system to illustrate the theoretical concepts that are developed throughout this book.

The screen display consists of a Title section, a Graph section, and a Track and Mark section. All three of these sections may be modified to meet the user's requirements. The Title section information is summarized in Table 1-3 and the channel designations and descriptions that are included in the trace fields are summarized in Table 1-4. There are nine trace fields illustrated in the Graph section. Note that the first and second trace fields are overlay groupings of the three phase-to-phase voltages and the three phase currents.

FREQUENCY, VOLTAGE, AND CURRENT 7

Table 1-3 Description of Title Section Information for the Digital Fault Recorder

ITEM	DESCRIPTION
Transcan	The name of the digital fault recorder that was used to provide this data
Graph #9	The power system analysis software that is provided with the Transcan digital fault recorder. It will accommodate 11 graphs for the display of such quantities as real and reactive power, rms values, harmonics, and other information.
1x	The time scale which is measured along the vertical axis may be adjusted to permit viewing data over longer or shorter intervals of time. The magnification factor shown for Graph #9 is 1x which is the default value of "one time" magnification.
3024 Hz	The sampling rate. The magnitudes of the analog quantities are sampled at a rate of 3024 times every second. The measured values are then stored in the memory of the digital fault recorder.
Record: SDGEPOWR.XP1	The data acquired for this event are stored in a file named SDGEPOWR.XP1. This file may be retrieved at any time to conduct further analyis of the measured data.
157 Days, 12:38:09	The date and time of the event. The number 157 represents the 157th day of the year. The event occurred at 12:38:09 which is the 12th hour, 38th minute, and 9th second of the day.

CHAPTER 1

Table 1-4 Channel Designations and Descriptions

CHANNEL DESIGNATIONS	DESCRIPTION
69-V_{ab}	Instantaneous voltage between a-phase and b-phase
69-V_{bc}	Instantaneous voltage between b-phase and c-phase
69-V_{ca}	Instantaneous voltage between c-phase and a-phase
696-I_a	a-phase current
696-I_b	b-phase current
696-I_c	c-phase current
69-V_{ca}-f	Frequency deviation for V_{ca}

The data items that were selected for this graph include the three phase-to-phase voltages, the three phase currents, and the frequency deviation associated with the V_{ca} voltage. Note that the frequency is a three-phase phenomenon and it is therefore only necessary to show the frequency deviation for one phase. Scale bars are shown at 31 milliseconds (ms) and 89 ms which can be read at the rightmost column of the graph.

The scale bar for each trace field has a center value of zero with five graduations in the negative direction and five graduations in the positive direction. This may vary for quantities such as root-mean-square (rms) values which are always positive. The scales for the trace fields in kilovolts per division, amperes per division, and millihertz per division may be shown in the Title section but have been omitted in this scan for clarity.

The Track and Mark section appears at the bottom of the screen display. Ten markers numbered 0 through 9 are available to view the values of data at specific points in the output. Marker 1 is shown at the left of the Track and Mark section. The data identified here is for

FREQUENCY, VOLTAGE, AND CURRENT 9

the trace field 69-V_{ca}-f at 44.0 ms and is also identified by Marker 1 at that point of the trace in the graph section. The measured frequency deviation at that point is -18.1 millihertz (mHz) which results in an actual frequency of

$$\begin{aligned} f &= f + \Delta f \\ &= 60 + (-0.0181) \\ &= 59.9819 \text{ Hz} \end{aligned} \quad (1\text{-}12)$$

An electric utility may declare an alert state when the frequency has declined to 59.95 Hz. An actual frequency which is less than the scheduled frequency may be indicative of an inadequate level of generation to meet the load requirements. The deviation measured in this event would be considered acceptable for most electric power systems.

The value associated with the screen cursor is shown at the right of the Track and Mark section. The data identified here is for the trace field 696-I_a at 44.0 ms which shows the instantaneous value of the a-phase current to be 153.7 amperes (A). The cursor is identified in the graph section by a plus (+) sign on the trace. The cursor may be moved to any trace and to any point of time in the record to identify the value of the desired quantity at that point.

1.3 SINUSOIDAL VOLTAGE AND CURRENT

Sinusoidal and cosinusoidal terms are related by the following trigonometric identities:

$$\cos\theta = \sin\left(\theta + \frac{\pi}{2}\right) \quad (1\text{-}13)$$

$$\sin\theta = \cos\left(\theta - \frac{\pi}{2}\right) \quad (1\text{-}14)$$

The reader is introduced to these identities and those in the following sections in order to define some of the fundamental relationships that are necessary for power systems analysis.

Sinusoidal and cosinusoidal waveforms for a single cycle are illustrated in Fig 1-3. The period for these waveforms is 16.67 ms which corresponds to the scheduled power system frequency of 60 Hz. This is the fundamental frequency for the voltages and currents of the electric power system. Multiples of the fundamental frequency are known as harmonics which may also be present as the result of abnormal conditions on the system. The second harmonic is 120 Hz and is the predominant harmonic in transformer inrush currents. The third harmonic is 180 Hz and is the predominant harmonic in system neutral currents. The evaluation of harmonic voltages and currents is important to the efficient operation of the system and will be reviewed in detail in subsequent sections.

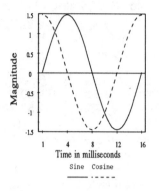

Figure 1-3 Sinusoidal and cosinusoidal waveforms.

1.4 RMS VALUES OF SINUSOIDAL VOLTAGE AND CURRENT

The significance of the rms values of voltage and current is based on the equivalence between these values and the direct current (dc) values that would yield the same power transfer in a dc circuit. We will prove this in Chap. 2, Power. The square of the voltage function $v(t)$ is represented as $v^2(t)$. The rms value of the voltage function can be stated in terms of the following three-step process.

The first step is to compute the sum of the squares for the voltage function V_{SS}.

$$V_{SS} = \int_0^\tau v^2(t)dt \qquad (1\text{-}15)$$

The second step is to determine the mean square voltage V_{MS}. This is accomplished by dividing the sum-of-squares voltage V_{SS} by the period τ.

FREQUENCY, VOLTAGE, AND CURRENT

$$V_{MS} = \frac{V_{SS}}{\tau} = \frac{1}{\tau}\int_0^\tau v^2(t)dt \qquad (1\text{-}16)$$

The third step is to determine the rms voltage V_{RMS}. This is accomplished by computing the square root of the mean square voltage as follows:

$$V_{RMS} = \sqrt{V_{MS}} = \sqrt{\frac{1}{\tau}\int_0^\tau v^2(t)dt} \qquad (1\text{-}17)$$

We may now compute the rms value associated with an actual voltage for the electric power system. Consider the following cosinusoidal voltage:

$$v(t) = V_m \cos(\omega t + \theta_V)$$

where V_m is the maximum or peak value of the voltage in volts, ω is the radian frequency in radians per second, t is the time in seconds, and θ_V is the voltage angle in radians.

The square of the cosinusoidal voltage is illustrated in Fig 1-4 and is given by the following expression:

$$v^2 = [V_m \cos(\omega t + \theta_V)]^2$$
$$= V_m^2 \cos^2(\omega t + \theta_V)$$

Note that the squared or second-degree cosine function is positive over the interval of a complete cycle. This characteristic yields rms values which are always positive.

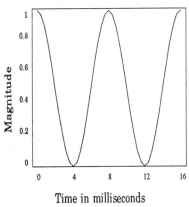

Figure 1-4 Squared cosine wave.

CHAPTER 1

The sum-of-squares voltage V_{SS} is computed as follows:

$$V_{SS} = \int_0^\tau [V_m \cos(\omega t + \theta_V)]^2 dt \qquad (1\text{-}19)$$

The expansion of the squared term yields

$$\begin{aligned}V_{SS} &= \int_0^\tau V_m^2 \cos^2(\omega t + \theta_V) dt \\ &= V_m^2 \int_0^\tau \cos^2(\omega t + \theta_V) dt\end{aligned} \qquad (1\text{-}20)$$

The *power relation* expressed in the squared cosinusoidal term may be resolved via the following trigonometric identity:

$$\cos^2 u = \frac{1}{2}(1 + \cos 2u) \qquad (1\text{-}21)$$

The expression is then rewritten as follows:

$$\begin{aligned}V_{SS} &= V_m^2 \int_0^\tau \frac{1}{2}[1 + \cos 2(\omega t + \theta_V)] dt \\ &= \frac{V_m^2}{2}\left[\int_0^\tau dt + \int_0^\tau \cos 2(\omega t + \theta_V) dt\right]\end{aligned} \qquad (1\text{-}22)$$

The solution of the integral contains the following forms:

$$\int du = u + C \qquad \int \cos v\, dv = \sin v + C$$

where $du = dt$, $\cos v\, dv = \cos 2(\omega t + \theta_V) d2(\omega t + \theta_V)$, and C is the constant of integration. Manipulating the integral to achieve the appropriate form yields

$$V_{SS} = \frac{V_m^2}{2}\left[\int_0^\tau dt + \frac{1}{2\omega}\int_0^\tau \cos 2(\omega t + \theta_V)(2\omega dt)\right] \qquad (1\text{-}23)$$

FREQUENCY, VOLTAGE, AND CURRENT

The solution of the integral yields

$$V_{ss} = \frac{V_m^2}{2}\left[\int_0^\tau dt + \frac{1}{2\omega}\int_0^\tau \cos 2(\omega t + \theta_v)(2\omega dt)\right]$$
$$= \frac{V_m^2}{2}\left[t\Big|_0^\tau + \frac{1}{2\omega}\sin(2\omega t + 2\theta_v)\Big|_0^\tau\right] \qquad (1\text{-}24)$$
$$= \frac{V_m^2}{2}\left[\tau + \frac{1}{2\omega}[\sin(2\omega\tau + 2\theta_v) - \sin(2\theta_v)]\right]$$

The *angle-sum relation* in the term $\sin(2\omega\tau + 2\theta_v)$ may be resolved by the following trigonometric identity:

$$\sin(\alpha + \beta) = \sin\alpha\cos\beta + \cos\alpha\sin\beta \qquad (1\text{-}25)$$

where $\alpha = 2\omega\tau$ and $\beta = 2\theta_v$. The following relationships are useful to the resolution of this expression.

$$f = \frac{1}{\tau} \qquad \omega = 2\pi f \qquad \omega\tau = \frac{2\pi}{\tau}\cdot\tau = 2\pi \qquad (1\text{-}26)$$

The angle-sum term may then be represented as follows:

$$\begin{aligned}\sin(2\omega\tau + 2\theta_v) &= \sin(2\omega\tau)\cos(2\theta_v) + \cos(2\omega\tau)\sin(2\theta_v) \\ &= \sin(4\pi)\cos(2\theta_v) + \cos(4\pi)\sin(2\theta_v) \\ &= (0)\cos(2\theta_v) + (1)\sin(2\theta_v) \\ &= \sin 2\theta_v\end{aligned} \qquad (1\text{-}27)$$

The solution of the sum-of-squares voltage therefore yields

$$\begin{aligned}V_{ss} &= \frac{V_m^2}{2}\left[\tau + \frac{1}{2\omega}[\sin(2\omega\tau + 2\theta_v) - \sin(2\theta_v)]\right] \\ &= \frac{V_m^2}{2}\left[\tau + \frac{1}{2\omega}(\sin 2\theta_v - \sin 2\theta_v)\right] \\ &= \frac{V_m^2 \tau}{2}\end{aligned} \qquad (1\text{-}28)$$

The mean square voltage is computed by dividing the sum-of-squares voltage by the period τ as follows:

$$V_{MS} = \frac{V_{SS}}{\tau} = \frac{V_m^2 \tau}{2\tau} = \frac{V_m^2}{2} \qquad (1\text{-}29)$$

The rms voltage is then found as the square root of the mean-square-voltage as follows:

$$V_{RMS} = \sqrt{\frac{V_m^2}{2}} = \frac{V_m}{\sqrt{2}} \qquad (1\text{-}30)$$

The rms voltage may therefore be expressed as follows:

$$V_{RMS} = 0.707 V_m \qquad (1\text{-}31)$$

The derivation of the rms value of a cosinusoidal current is performed in the same manner as for a cosinusoidal voltage and results in the following expression:

$$I_{RMS} = 0.707 I_m \qquad (1\text{-}32)$$

The symbols V and I will be used to designate rms voltage and current, respectively, throughout this text.

1.5 MEASUREMENT OF RMS QUANTITIES

The instantaneous and rms values of the phase-to-phase voltages and phase currents for a 69-kV subtransmission line are shown in Fig 1-5. The marker numbers which appear on each trace may be correlated with the corresponding values listed in the Title section to identify each trace.

The first trace field includes the instantaneous values of the three phase-to-phase voltages V_{ab}, V_{bc}, and V_{ca}. The second trace field includes the instantaneous values of the three phase currents I_a, I_b, and I_c. The remaining traces are rms values of the phase-to-phase voltages and the phase currents which are computed from the data of the previous 10 cycles. The values shown in the graph reflect the normal values of voltage and current that existed just prior to a disturbance on the power system. This is known as prefault data. We will review data which were obtained after the occurrence of the fault, e.g., postfault data, in later chapters.

FREQUENCY, VOLTAGE, AND CURRENT

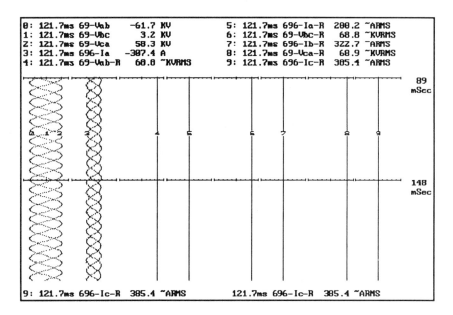

Figure 1-5 Instantaneous and rms values of three-phase voltages and currents. (*Courtesy of Mehta Tech, Inc.*)

1.6 DIGITAL COMPUTATION OF RMS QUANTITIES

The rms value of voltage and current illustrated in the above figures were obtained by measuring the magnitudes of the analog values at a sampling rate of 3024 Hz or 3024 samples per second. The scheduled power system frequency of 60 Hz yields the following number of samples per cycle:

$$\frac{3024 \text{ samples}}{\text{s}} \cdot \frac{1 \text{ s}}{60 \text{ cycles}} = 50.4 \frac{\text{samples}}{\text{cycle}} \qquad (1\text{-}33)$$

The digital fault recorder determines the rms value on the basis of the previous 10 cycles or 504 samples. The sampled values are then squared and summed in a register designated as SS or sum of squares. The sum of squares is then divided by the number of samples (504) to obtain the MS or mean square value. The square root of this value

CHAPTER 1

then yields the root of the mean square or rms value of voltage[2] or current. The following data were generated by a simple spreadsheet macro which performed the following steps:

1. Generated 10 cycles of a sine wave with a frequency of 60 Hz and a peak magnitude of 1.
2. Sampled the magnitude of the sine wave at 3024 Hz to yield 504 samples over ten cycles.
3. Squared the value determined for each sample.
4. Summed the squared values to obtain the sum of squares SS.
5. Divided the sum of squares by the number of samples n (504) to obtain the mean square MS.
6. Computed the square root of the mean square to obtain the rms value of 0.707 which is consistent with our results from the previous section.

Table 1-5 Digital RMS Quantities

NUMBER OF SAMPLES (n)	SUM OF SQUARES (SS)	MEAN SQUARE (MS)	ROOT MEAN SQUARE (RMS)
504	252	0.5	0.707

The digital computation of the rms value is summarized in Table 1-5. The worksheet data listed in this table were excerpted from a personal computer-based spreadsheet program. These programs are often useful for applications such as the reporting of tabular information, calculations, and graphs. Much of the tabular data for this manual was generated in such a manner.

---

[2]The unit volt (V) as a measure of voltage is named in honor of the Italian physicist Count Alessandro Volta (1745-1827) who demonstrated the production of electric current by the action of two plates of different metals in an electrolyte.

FREQUENCY, VOLTAGE, AND CURRENT   17

1.7 AVERAGE VALUE OF SINUSOIDAL VOLTAGE AND CURRENT

The general expression for the average value of a voltage signal is given as follows:

$$V_{AVE} = \frac{1}{\tau}\int_0^\tau v(t)dt \qquad (1\text{-}34)$$

Consider the following cosinusoidal voltage:

$$v(t) = V_m \cos(\omega t + \theta_V) \qquad (1\text{-}35)$$

The expression for the average value of the cosinusoidal voltage is

$$V_{AVE} = \frac{1}{\tau}\int_0^\tau V_m \cos(\omega t + \theta_V)dt \qquad (1\text{-}36)$$

The solution of the integral is of the form

$$\int \cos u\, du = \sin u + C \qquad (1\text{-}37)$$

where $\cos u\, du = \cos(\omega t + \theta_V)(\omega dt)$ and C is the constant of integration. Manipulating the equation to achieve to appropriate form and solving yields

$$\begin{aligned}V_{AVE} &= \frac{V_m}{\omega\tau}\int_0^\tau \cos(\omega t + \theta_V)(\omega dt) \\ &= \frac{V_m}{\omega\tau}\sin(\omega t + \theta_V)\Big|_0^\tau \\ &= \frac{V_m}{\omega\tau}[\sin(\omega\tau + \theta_V) - \sin(\theta_V)]\end{aligned} \qquad (1\text{-}38)$$

The angle-sum relation in the term $\sin(\omega\tau + \theta_V)$ may be resolved by the trigonometric identity of (1-25) which is restated for convenience as follows:

$$\sin(\alpha + \beta) = \sin\alpha\cos\beta + \cos\alpha\sin\beta$$

where $\alpha = \omega\tau$ and $\beta = \theta_V$. The relationships of (1-26) are also useful to the solution of the integral and are restated for convenience.

CHAPTER 1

$$f=\frac{1}{\tau} \quad \omega=2\pi f=\frac{2\pi}{\tau} \quad \omega\tau=\frac{2\pi}{\tau}\cdot\tau=2\pi$$

The term may then be resolved as follows:

$$\begin{aligned}\sin(\omega\tau+\theta_V)&=\sin(\omega\tau)\cos(\theta_V)+\cos(\omega\tau)\sin(\theta_V)\\&=\sin(2\pi)\cos(\theta_V)+\cos(2\pi)\sin(\theta_V)\\&=(0)\cos(\theta_V)+(1)\sin(\theta_V)\\&=\sin\theta_V\end{aligned} \quad (1\text{-}39)$$

The solution of the integral is then

$$\begin{aligned}V_{AVE}&=\frac{V_m}{\omega\tau}[\sin(\omega\tau+\theta_V)-\sin(\theta_V)]\\&=\frac{V_m}{2\pi}[\sin\theta_V-\sin\theta_V]\\&=0\end{aligned} \quad (1\text{-}40)$$

The average value of a cosinusoidal voltage over the period of a complete cycle is therefore equal to zero. This result occurs because of the symmetry of the cosinusoidal function about the zero axis over a complete cycle. Note also that this result is independent of the voltage angle θ_V. Consider instead the average value of the cosinusoidal function during the positive half-cycle from $\omega t=-\pi/2$ to $\omega t=\pi/2$.

$$\begin{aligned}V_{AVE}&=\frac{V_m}{\omega(t_2-t_1)}[\sin(\omega t+\theta_V)]\Big|_{t_1}^{t_2}\\&=\frac{V_m}{\omega\left(\frac{\pi}{2}-\left(-\frac{\pi}{2}\right)\right)}\left[\sin\left(\frac{\pi}{2}+\theta_V\right)-\sin\left(-\frac{\pi}{2}+\theta_V\right)\right]\end{aligned} \quad (1\text{-}41)$$

Invoking the trigonometric identity for angle-sum relations cited previously yields as follows:

$$V_{AVE}=\frac{V_m}{\pi}\left[\sin\left(\frac{\pi}{2}\right)\cos(\theta_V)+\cos\left(\frac{\pi}{2}\right)\sin(\theta_V)-\sin\left(-\frac{\pi}{2}\right)\cos(\theta_V)-\cos\left(-\frac{\pi}{2}\right)\sin(\theta_V)\right]$$

FREQUENCY, VOLTAGE, AND CURRENT 19

Resolving terms yields as follows:

$$V_{AVE} = \frac{V_m}{\pi}\left[(1)\cos(\theta_V)+(0)\sin(\theta_V)-(-1)\cos(\theta_V)-\cos\left(-\frac{\pi}{2}\right)\sin(\theta_V)\right]$$

$$= \frac{V_m}{\pi}[\cos(\theta_V)+\cos(\theta_V)]$$

$$= \frac{2V_m}{\pi}\cos(\theta_V)$$

Assuming a voltage angle of $\theta_V = 0$ yields

$$V_{AVE} = \frac{2V_m}{\pi}\cos(0) = \frac{2V_m}{\pi} \quad (1\text{-}42)$$

The average value of a sinusoidal voltage over the period of one-half cycle is

$$V_{AVE} = 0.637 V_m \quad (1\text{-}43)$$

The derivation of the average value of a cosinusoidal current over the same time interval is performed in the same manner as for a cosinusoidal voltage and results in the following expression:

$$I_{AVE} = 0.637 I_m \quad (1\text{-}44)$$

1.8 VOLTAGE AND CURRENT MAGNITUDES

The quantities associated with cosinusoidal voltage and current waveforms that were analyzed in the previous sections are illustrated in graphical form in Fig. 1-6. The rms value represents the effective value of the waveform and will be used throughout this book to represent voltage and current unless otherwise stated. The rms values are used for steady-state quantities to determine the operating ratings.

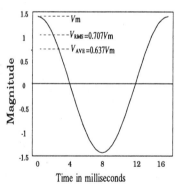

Figure 1-6 Cosinusoidal magnitudes.

CHAPTER 1

Table 1-6 Voltage and Current Conventions

VALUE	VOLTAGE	CURRENT
Maximum or peak	V_m	I_m
Root mean square or effective	$V_{RMS}=0.707V_m=V$	$I_{RMS}=0.707I_m=I$
Average	$V_{AVE}=0.637V_m$	$I_{AVE}=0.637I_m$
Instantaneous	$v=V_m\cos(\omega t+\theta_V)$	$i=I_m\cos(\omega t+\theta_I)$
Angle	θ_V	θ_I

Voltage on the bulk power system is generally measured in volts or thousands of volts (kilovolts, kV). Current is measured in amperes or thousands of amperes (kiloamperes, kA). The quantities associated with sinusoidal voltage and current waveforms are summarized in tabular form in Table 1-6. It should be noted that the consistent use of conventions for power systems quantities is of the utmost importance in the analysis of electric power systems. In the next section we will discuss a representation for voltage and current which is useful for the analysis of time-varying quantities.

1.9 VOLTAGE AND CURRENT PHASORS

A *vector* is a quantity which represents a magnitude and an angle. A *phasor* is a quantity which rotates at a radian frequency ω and has an angle θ at time $t=0$. A sinusoidal voltage is a phasor quantity since it has a magnitude V_m, a radian frequency ω, and a phase angle θ_V at time $t=0$; similarly, a sinusoidal current is a phasor quantity since it has a magnitude I_m, a radian frequency ω, and a phase angle θ_I at time $t=0$.

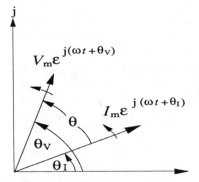

Figure 1-7 Rotating voltage and current phasors.

FREQUENCY, VOLTAGE, AND CURRENT 21

The term *phasor* is used to define electrical quantities which are sinusoidal in nature in order to distinguish them from space vectors which are used to define space coordinates. We will use the term *phasor* exclusively in this text to simplify matters. Phasors will be represented in upper case bold italics such as ***V*** or ***I***, and phasor magnitudes will be represented in upper case normal italics such as V or I.

Consider Euler's theorem which is stated as follows:

$$\varepsilon^{j\theta} = \cos\theta + j\sin\theta \qquad (1\text{-}45)$$

The conjugate of Euler's theorem is stated as follows:

$$\varepsilon^{-j\theta} = \cos\theta - j\sin\theta \qquad (1\text{-}46)$$

The left-hand side of the equation is expressed in exponential form and the right-hand side is expressed in rectangular form. The real and imaginary components of this equation may be stated as follows:

$$\Re\varepsilon^{j\theta} = \cos\theta \qquad (1\text{-}47)$$

$$\Im\varepsilon^{j\theta} = \sin\theta \qquad (1\text{-}48)$$

Euler's theorem may be rewritten to obtain the following expressions:

$$\cos\theta = \frac{1}{2}(\varepsilon^{j\theta} + \varepsilon^{-j\theta}) \qquad (1\text{-}49)$$

$$\sin\theta = \frac{1}{2}(\varepsilon^{j\theta} - \varepsilon^{-j\theta}) \qquad (1\text{-}50)$$

Cosinusoidal voltages and currents may be represented as follows:

$$v = V_m \cos(\omega t + \theta_V) \qquad (1\text{-}51)$$

$$i = I_m \cos(\omega t + \theta_I) \qquad (1\text{-}52)$$

CHAPTER 1

The exponential representations for these quantities are

$$v = \Re V_m \varepsilon^{j(\omega t + \theta_V)} = \Re V_m \varepsilon^{j\omega t} \varepsilon^{j\theta_V} \qquad (1\text{-}53)$$

$$i = \Re I_m \varepsilon^{j(\omega t + \theta_I)} = \Re I_m \varepsilon^{j\omega t} \varepsilon^{j\theta_I} \qquad (1\text{-}54)$$

It is standard practice to use the rms values instead of the maximum values for the magnitudes of the phasors. The magnitudes of the voltage and current phasors are therefore given as follows:

$$|V| = \frac{V_m}{\sqrt{2}} = V \qquad (1\text{-}55)$$

$$|I| = \frac{I_m}{\sqrt{2}} = I \qquad (1\text{-}56)$$

In addition, we will establish reference phasors for voltage and current at $t=0$ which results in the following exponentials:

$$\varepsilon^{j(\omega t + \theta_V)}\big|_{t=0} = \varepsilon^{j\theta_V} \qquad (1\text{-}57)$$

$$\varepsilon^{j(\omega t + \theta_I)}\big|_{t=0} = \varepsilon^{j\theta_I} \qquad (1\text{-}58)$$

The phasor notation for voltage and current may now be expressed as follows:

$$V = V\varepsilon^{j\theta_V} = V\angle\theta_V \qquad (1\text{-}59)$$

$$I = I\varepsilon^{j\theta_I} = I\angle\theta_I \qquad (1\text{-}60)$$

The apparent impedance at the terminals of a transmission line may be determined by the ratio of the phasor voltage to the phasor current as follows:

$$Z = \frac{V}{I} = \frac{V}{I}\angle(\theta_V - \theta_I) \qquad (1\text{-}61)$$

FREQUENCY, VOLTAGE, AND CURRENT 23

The conjugates of the phasors for voltage and current are expressed as follows:

$$\hat{V}=V\varepsilon^{-j\theta_v}=V\angle-\theta_V \qquad (1\text{-}62)$$

$$\hat{I}=I\varepsilon^{-j\theta_I}=I\angle-\theta_I \qquad (1\text{-}63)$$

The conjugate of the voltage phasor is obtained by reversing the direction of rotation from the positive or counterclockwise direction $j\omega t$ to the negative or clockwise direction $-j\omega t$ and by reversing the sign of the voltage angle from θ_V to $-\theta_V$; similarly, the conjugate of the current phasor is obtained by reversing the direction of rotation from the positive or counterclockwise direction $j\omega t$ to the negative or clockwise direction $-j\omega t$ and by reversing the sign of the current angle from θ_I to $-\theta_I$.

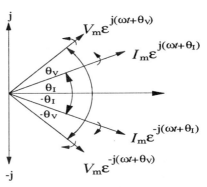

Figure 1-8 Complex conjugates of rotating voltage and current phasors.

1.10 ANGULAR RELATIONSHIPS
1.10.1 Electrical Length
The velocity of electromagnetic radiation in free space is

$$v=186{,}000 \ \frac{\text{mi}}{\text{s}} \qquad (1\text{-}64)$$

The scheduled power system frequency is

$$f=60 \ \text{cps} \qquad (1\text{-}65)$$

The wavelength of on complete cycle of a 60-Hz waveform is

$$\lambda=\frac{186{,}000 \ \text{mi}}{\text{s}} \cdot \frac{1 \ \text{s}}{60 \ \text{cycles}}=3100 \ \frac{\text{mi}}{\text{cycle}} \qquad (1\text{-}66)$$

It is unreasonable to consider a power transmission line of such a length. The electrical length of a 1-MHz sine wave, however, is computed in the same manner and found to be approximately 982 feet (ft). A quarter-wavelength antenna for radio frequency applications at 1 MHz would therefore be approximately 246 ft in length. A complete cycle traverses 360° or 2π rad.

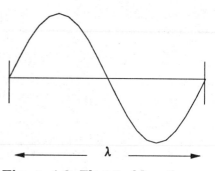

Figure 1-9 Electrical length.

The linear distance per degree or radian for the power system frequency of 60 Hz may be computed as follows:

$$\frac{360 \text{ degrees}}{3100 \text{ mi}} = 0.116 \frac{\text{degrees}}{\text{mi}} \qquad (1\text{-}67)$$

$$\frac{2\pi \text{ rad}}{3100 \text{ mi}} = 0.002 \frac{\text{rad}}{\text{mi}} \qquad (1\text{-}68)$$

The electrical length of a transmission line is the number of degrees or radians of a 60-Hz voltage angle or current angle that will appear between the terminals of the line. Most transmission lines in North America are less than 100 mi in length. The electrical length of a 100-mi transmission line is

$$100 \text{ mi} \cdot 0.116 \frac{\text{degrees}}{\text{mi}} = 11.6 \text{ degrees} \qquad (1\text{-}69)$$

$$100 \text{ mi} \cdot 0.002 \frac{\text{rad}}{\text{mi}} = 0.2 \text{ rad} \qquad (1\text{-}70)$$

The analysis of voltage and current on the electric power transmission systems is simplified for electrically short lines by neglecting the difference in phase between buses. This approximation yields results that are not significantly different from those that are obtained from exact calculations. This is not true of electrically long circuits which can have a substantial angular difference between terminals.

1.10.2 Phase Angles

The phase angle θ is the difference between the voltage angle θ_V and the current angle θ_I at a given point of the system where $\theta = \theta_V - \theta_I$. The voltage phasor is assumed to be the reference quantity $\theta_V = 0$ in problems involving the phase angle between voltage and current.

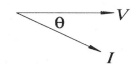

Figure 1-10 Phase angle.

1.10.3 Voltage Angles

The voltage angle θ_V represents the difference in phase between the voltage at a reference bus for which the voltage angle is assumed to be zero at $t=0$ and the voltage at the bus under consideration which may be at some other angle. We will see in subsequent sections that phase shifts occur in power systems as the result of load flow and wye-delta transformations where at some arbitrary time the voltages at various buses will be shifted in phase by some angle.

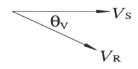

Figure 1-11 Voltage angle.

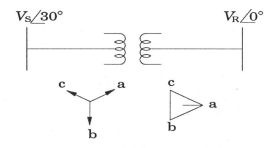

Figure 1-12 Sending and receiving bus voltages.

1.10.4 Current Angles

The current I_S at the sending terminal of a transmission line may be compared to the current I_R at the receiving end to determine if an internal fault exists on the line. The current flow at the line terminals reverses direction every half-cycle so that the relative direction of fault current must be determined by the phase difference between terminals.

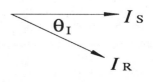

Figure 1-13 Current angle.

The phase relationship for normal operating conditions and for faults that are external to the transmission line is such that the current which flows into the sending terminal of the transmission line is essentially in phase with the current that flows out of the receiving terminal. This is based on the assumption of an electrically short line.

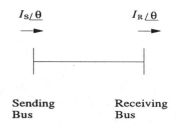

Figure 1-14 Normal bus currents.

An internal transmission line fault will result in a reversal of the current at one terminal so that the net flow is into the transmission line. The detection of transmission line faults is therefore often accomplished by comparison of the current angles at either terminal. This form of transmissoin line protective relaying is referred to as *phase comparison*.

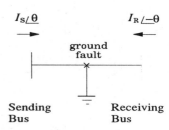

Figure 1-15 Faulted bus currents.

FREQUENCY, VOLTAGE, AND CURRENT 27

1.11 THREE-PHASE VOLTAGE AND CURRENT

The phase voltages V_A, V_B, V_C and the phase-to-phase voltages V_{AB}, V_{BC}, V_{CA} in a balanced three-phase system are related by the factor $\sqrt{3}$. This factor is derived from the law of cosines which may be used to determine the length of a side of a triangle when the length of the other two sides and the corresponding angle are known. The law of cosines is stated as follows:

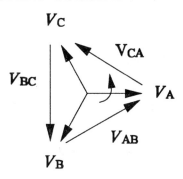

$$c^2 = a_2 + b_2 - 2ab\cos\theta \qquad (1\text{-}71)$$

where c is the magnitude of the unknown side of the triangle, a and b are the magnitudes of the known sides, and θ is the angle between the known sides. The magnitudes of the three phase voltages in a balanced system are equal and may be expressed as V_ϕ; similarly, the magnitudes of the three phase-to-phase voltages in a balanced system are equal and may be expressed as $V_{\phi\phi}$.

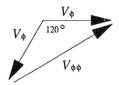 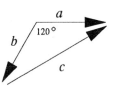

The phase voltages may be substituted into the law of cosines as follows:

$$a = b = V_\phi \qquad (1\text{-}72)$$

The corresponding phase-to-phase voltage may be substituted into the law of cosines as follows:

$$c = V_{\phi\phi} \qquad (1\text{-}73)$$

CHAPTER 1

The angle between the phase voltages may be substituted into the law of cosines as follows:

$$\theta = 120° \tag{1-74}$$

Solution by the law of cosines yields

$$\begin{aligned}V_{\phi\phi} &= V_\phi^2 + V_\phi^2 - 2V_\phi^2 \cos(120°) \\ &= V_\phi^2 + V_\phi^2 - 2V_\phi^2(-0.5) \\ &= 3V_\phi^2 \end{aligned} \tag{1-75}$$

The relationship between V_ϕ and $V_{\phi\phi}$ may be stated as follows:

$$V_{\phi\phi} = \sqrt{3}\, V_\phi \tag{1-76}$$

The various forms of the three-phase voltage phasors are summarized in Table 1-7. The polar and exponential representations are most useful for phasor operations involving multiplication and division. The rectangular representation is most useful for phasor operations involving addition and subtraction. The computation of voltage as the product of current and impedance may be performed using polar or exponential representations.

Table 1-7 Forms of Representation for Three-phase Sinusoidal Voltages

ϕ	VOLTAGE MAGNITUDE	VOLTAGE ANGLE	POLAR FORM	EXPONENTIAL FORM	RECTANGULAR FORM
A	V_ϕ	0°	$V_\phi \angle 0°$	$V_\phi \varepsilon^{+j0°}$	$V_\phi(1+j0)$
B	V_ϕ	-120°	$V_\phi \angle -120°$	$V_\phi \varepsilon^{-j120°}$	$V_\phi\left(-\dfrac{1}{2} - j\dfrac{\sqrt{3}}{2}\right)$
C	V_ϕ	+120°	$V_\phi \angle +120°$	$V_\phi \varepsilon^{+j120°}$	$V_\phi\left(-\dfrac{1}{2} + j\dfrac{\sqrt{3}}{2}\right)$

FREQUENCY, VOLTAGE, AND CURRENT

$$V = IZ = I\angle\theta_I \; Z\angle\theta_Z = IZ\angle(\theta_I + \theta_Z) \quad (1\text{-}77)$$

$$V = IZ = I\varepsilon^{j\theta_I} Z\varepsilon^{j\theta_Z} = IZ\varepsilon^{j(\theta_I + \theta_Z)} \quad (1\text{-}78)$$

The computation of impedance as the ratio of voltage and current may be performed using polar and exponential representations as follows:

$$Z = \frac{V}{I} = \frac{V\angle\theta_V}{I\angle\theta_I} = \frac{V}{I}\angle\theta_V - \theta_I \quad (1\text{-}79)$$

$$Z = \frac{V}{I} = \frac{V\varepsilon^{j\theta_V}}{I\varepsilon^{j\theta_I}} = \frac{V}{I}\varepsilon^{j(\theta_V - \theta_I)} \quad (1\text{-}80)$$

The computation of a source voltage as the sum of a circuit voltage and a load voltage may be performed using the rectangular representation as follows:

$$V_S = V_C + V_L$$
$$= (V_C\cos\theta_C + V_L\cos\theta_L) + j(V_C\sin\theta_C + V_L\sin\theta_L) \quad (1\text{-}81)$$

Table 1-8 Phasor Components for Phase-to-Phase Voltages

PHASE-TO-PHASE VOLTAGES	RELATIONSHIP	PHASOR DIAGRAM
V_{AB}	$V_{AB} = V_A - V_B$	
V_{BC}	$V_{BC} = V_B - V_C$	
V_{CA}	$V_{CA} = V_C - V_A$	

The phasor components for the phase-to-phase voltages are summarized in Table 1-8. We conclude this discussion with the thought that proficiency in the use of phasor notation is a most important requirement for power systems analysis.

1.12 POWER SYSTEM HARMONICS

The Fourier series is a convenient method to represent the voltage and current on transmission lines when frequencies other than the fundamental or scheduled power system frequency of 60 Hz are present. The purpose of this discussion is to demonstrate that the voltage and current phasors of the fundamental frequency rotate in a positive phase sequence, those of the second harmonic frequency rotate in a negative phase sequence, and those of the third harmonic frequency rotate in phase. The exponential form of the voltage phasor is expressed as

$$v(t) = V \varepsilon^{j\omega t} \tag{1-82}$$

where V is the magnitude of the voltage phasor and $j\omega t$ is the argument. The Fourier series for the general expression $v(t)$ may be written as

$$v(t) = \sum_{n=-\infty}^{n=\infty} V_n \varepsilon^{jn\omega_b t} = V_1 \varepsilon^{j\omega_b t} + V_2 \varepsilon^{j2\omega_b t} + V_3 \varepsilon^{j3\omega_b t} + \cdots \tag{1-83}$$

Consider a three-phase system with the phase voltages expressed in exponential representation for the fundamental frequency components.

$$v_A(t) = V_A \varepsilon^{j\omega_b t} \qquad v_B(t) = V_B \varepsilon^{j(\omega_b t - 120°)} \qquad v_C(t) = V_C \varepsilon^{j(\omega_b t + 120°)}$$

Extending the terms of each of the three phases to the third harmonic yields

$$\begin{aligned}
v_A(t) &= V_{A1} \varepsilon^{j\omega_b t} + V_{A2} \varepsilon^{j2\omega_b t} + V_{A3} \varepsilon^{j3\omega_b t} + \cdots \\
v_B(t) &= V_{B1} \varepsilon^{j(\omega_b t - 120°)} + V_{B2} \varepsilon^{j(2\omega_b t - 240°)} + V_{B3} \varepsilon^{j(3\omega_b t - 360°)} + \cdots \\
v_C(t) &= V_{C1} \varepsilon^{j(\omega_b t + 120°)} + V_{C2} \varepsilon^{j(2\omega_b t + 240°)} + V_{C3} \varepsilon^{j(3\omega_b t + 360°)} + \cdots
\end{aligned} \tag{1-84}$$

FREQUENCY, VOLTAGE, AND CURRENT

Table 1-9 Three-Phase Harmonics

PHASE	FUNDAMENTAL FREQUENCY	SECOND HARMONIC	THIRD HARMONIC
A	0°	0°	0°
B	-120°	-240°	-360°
C	+120°	+240°	+360°

The arguments of the phase components of the fundamental frequency and the second and third harmonic frequencies are summarized in Table 1-9. It can be seen from the table that

1. The fundamental frequency components rotate in a positive sequence as A-phase (0°), B-phase (-120°), and C-phase (+120°).

2. The second harmonic components rotate in a negative sequence as A-phase (0°), B-phase (+120°), and C-phase (-120°).

3. The third harmonic components rotate in phase as A-phase (0°), B-phase (0°), and C-phase (0°).

This relationship repeats itself for the fourth, fifth, and sixth harmonics, again for the seventh, eighth, and ninth harmonics, and again thereafter. The significant consideration here is that, for V_A, V_B, and V_C, the fundamental and second harmonic phase components are balanced and sum to zero whereas the third harmonic phase components are in phase and flow through the neutrals of the power system.

It is for this reason that delta windings are used on power system transformers to trap the third harmonic components since the delta winding represents an open circuit to in-phase components. This becomes obvious when one considers that connecting the same voltage to the three terminals of a delta winding does not provide a return path to the source as in the case of a grounded wye in which the currents can return to the source through the grounded neutral terminal.

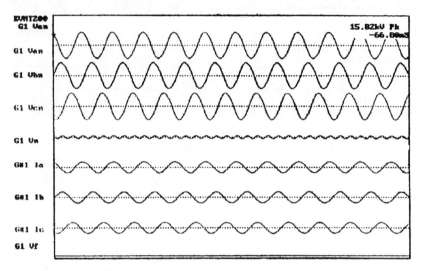

Figure 1-22 Digital fault recording of phase voltages, phase currents, and neutral voltage with third harmonic component.

The third harmonic component of the neutral voltage for a power system generator is illustrated in the digital fault recording of Fig. 1-16. The phase-to-neutral voltages for generator G_1 are V_{an}, V_{bn}, and V_{cn}, the phase currents are I_a, I_b, and I_c, and the neutral voltage is V_n. Note that there are three cycles of neutral voltage for each cycle of the phase voltages which are of fundamental frequency.

The neutral voltage was measured across the secondary winding of the neutral grounding transformer. The peak value of the phase-to-neutral voltage V_{an} is shown in the upper right corner of the graph as 15.82 kV. The rms value may be determined by the methods of this chapter to be 11.18 kV and the corresponding phase-to-phase voltage at the generator terminals is 19.37 kV.

CHAPTER 2

POWER

2.1 INSTANTANEOUS POWER

Consider the following cosinusoidal voltage:

$$v = V_m \cos(\omega t + \theta_V) \quad (2\text{-}1)$$

Now consider that the application of this voltage to the terminals of a transmission line gives rise to the following cosinusoidal current:

$$i = I_m \cos(\omega t + \theta_I) \quad (2\text{-}2)$$

The voltage phasor will be our reference in these analyses and the following angle relationships therefore exist:

$$\begin{aligned} \theta &= \theta_V - \theta_I \\ \theta_V &= 0 \\ \theta &= -\theta_I \\ \theta_I &= -\theta \end{aligned} \quad (2\text{-}3)$$

Based on these relationships, the voltage and current may be restated as follows:

$$v = V_m \cos\omega t \quad (2\text{-}4)$$

$$i = I_m \cos(\omega t - \theta) \quad (2\text{-}5)$$

The instantaneous power associated with this circuit is expressed as the product of the voltage and current as follows:

$$s = vi = (V_m \cos\omega t)(I_m \cos(\omega t - \theta)) = V_m I_m \cos\omega t \cos(\omega t - \theta) \quad (2\text{-}6)$$

The *function-product* relation in the expression $\cos\omega t \cos(\omega t - \theta)$ can be resolved by use of a trigonometric identity of the form $\cos\alpha\cos\beta$ where $\alpha = 2\omega t$ and $\beta = \theta$.

CHAPTER 2

$$\cos\alpha\cos\beta = \frac{1}{2}[\cos(\alpha-\beta)+\cos(\alpha+\beta)]$$

$$\cos\omega t\cos(\omega t-\theta) = \frac{1}{2}[\cos(\omega t-\omega t+\theta)+\cos(\omega t+\omega t-\theta)] \quad (2\text{-}7)$$

$$= \frac{1}{2}[\cos\theta+\cos(2\omega t-\theta)]$$

The *angle-difference* relation in the expression $\cos(2\omega t\text{-}\theta)$ can be further resolved by the use of the trigonometric identity of the form $\cos(\alpha\text{-}\beta)$ where $\alpha=2\omega t$ and $\beta=\theta$ as follows:

$$\cos(\alpha-\beta) = \cos\alpha\cos\beta + \sin\alpha\sin\beta$$
$$\cos(2\omega t-\theta) = \cos2\omega t\cos\theta + \sin2\omega t\sin\theta \quad (2\text{-}8)$$

The expression for the instantaneous power may therefore be stated as follows:

$$s = V_m I_m \left[\frac{1}{2}(\cos\theta + \cos2\omega t\cos\theta + \sin2\omega t\sin\theta)\right]$$
$$= \frac{V_m I_m}{2}\cos\theta(1+\cos2\omega t) + \frac{V_m I_m}{2}\sin\theta\sin2\omega t \quad (2\text{-}9)$$

We will now make use of the rms values for sinusoidal voltages and currents that were computed in Chap. 1 as follows:

$$V_{RMS} = \frac{V_m}{\sqrt{2}} \quad (2\text{-}10)$$

$$I_{RMS} = \frac{I_m}{\sqrt{2}} \quad (2\text{-}11)$$

The product of the voltage and current magnitudes may now be rewritten as follows:

$$\frac{V_m I_m}{2} = \frac{V_m}{\sqrt{2}}\frac{I_m}{\sqrt{2}} = VI \quad (2\text{-}12)$$

Now let us define the quantities that are listed in Table 2-1. These quantities represent the standard conventions that are used in the electric power industry.

POWER 35

Table 2-1 Summary of Real, Reactive and Apparent Power

SYMBOL	QUANTITY	UNITS	EXPRESSION
P	Real power	Watts (W)	$P = VI\cos\theta$
Q	Reactive power	Volt-amperes reactive (VAR)	$Q = VI\sin\theta$
S	Apparent power	Volt-amperes (VA)	$S = VI$

The instantaneous power may now be expressed in terms of the real and reactive power as follows:

$$s = P(1 + \cos 2\omega t) + Q \sin 2\omega t \qquad (2\text{-}13)$$

The product of the voltage in kilovolts and the current in kiloamperes yields the apparent power in megavolt-amperes (MVA), the real power in megawatts[3] (MW), and the reactive power in megavolt-amperes reactive (MVAR). These units are generally used in conjunction with the bulk electric power system. The power expressions may also be expressed in units of kilovolt-amperes (kVA), kilowatts (kW), and kilovolt-amperes reactive (kVAR).

2.2 REAL POWER (P)

The instantaneous value of the real power component is

$$p = P(1 + \cos 2\omega t) \qquad (2\text{-}14)$$

The real power component varies harmonically at twice the fundamental frequency about an average value P with a minimum value of zero and a maximum value of $2P$. It is significant to note that the real power P is computed on the basis of the rms values of voltage

---

[3]The unit watt is given in honor of the British inventor James Watt (1736-1819) who pioneered the quantification of the rate of change of energy.

and current and the cosine of the phase angle. The same real power would be delivered by dc magnitudes of voltage and current which are equal to the rms magnitudes. The quantity P therefore represents the average or dc equivalent value of power. Energy is transformed from various sources such as coal, oil, uranium, and hydraulic flow to electric energy via the prime movers of electric generating systems. Consider that the flow of water through the turbine of a hydroelectric power plant is 8000 cubic feet per second (ft³/s).

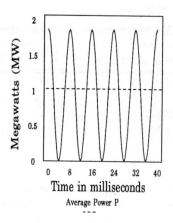

Figure 2-1 Real power.

$$Q = 8000 \text{ ft}^3/s \qquad (2\text{-}15)$$

The difference in elevation between the upper and lower reservoir is 300 feet. This is known as the available head.

$$h = 300 \text{ ft} \qquad (2\text{-}16)$$

The density of water is given as follows:

$$\gamma = 62.4 \text{ lb/ft}^3 \qquad (2\text{-}17)$$

The real power associated with the flow of water through the turbine is determined as follows:

$$P = Q\gamma h \qquad (2\text{-}18)$$

where P is the power in foot-pounds per second, Q is the flow rate in cubic feet per second, γ is the density of water in pounds per cubic foot, and h is the available head in feet.

POWER 37

The hydraulic power to the turbine is therefore

$$P = Q\gamma h$$
$$= (8000)(62.4)(300) \quad (2\text{-}19)$$
$$= 149{,}760{,}000 \text{ ft-lb/s}$$

A conversion factor may be applied to convert foot-pounds per second to megawatts as follows:

$$149{,}760{,}000 \text{ ft-lb/s} \cdot 0.000\,001\,356 = 203 \text{ MW} \quad (2\text{-}20)$$

The efficiency of the machine is less than 100 percent due to mechanical losses in the turbine and electrical losses in the generator. Assuming an efficiency of 90 percent yields

$$P_{out} = \eta P_{in}$$
$$= (0.9)(203) \quad (2\text{-}21)$$
$$= 183 \text{ MW}$$

This is the real power that is available at the terminals of the hydroelectric generator. The energy associated with this real power is then delivered to the industrial, commercial, and residential loads via the transmission, subtransmission, and distribution systems. Real power is consumed in the work that is performed by such loads as lighting, heating, ventilation, and air conditioning. Real power losses are incurred as the result of the flow of current through the resistance of the phase conductors of the electric power system.

2.3 REACTIVE POWER (Q)

The instantaneous value of the reactive power component is

$$q = Q\sin 2\omega t \quad (2\text{-}22)$$

The reactive power component varies harmonically at twice the fundamental frequency about an average value of zero with a minimum value of $-Q$ and a maximum value of $+Q$. The quantity Q therefore represents the power which is associated with the storage of energy in the magnetic field that results from the current flow through the inductance of the phase conductors and in the electric field that results from the application of the system voltage to the capacitance between the phase conductors and ground.

The power transfer capacity for a transmission system may be approximated with the following expression:

$$P = \frac{V_S V_R}{X} \sin\theta_V$$

where P is the three-phase power in megawatts, V_S and V_R are the phase-to-phase voltages in kilovolts at the sending and receiving terminals, X is the series reactance of the transmission line, and $\sin\theta_V$ is the sine of the voltage angle between the sending and receiving terminals. The power transfer capacity is proportional to the operating voltage which is in turn dependent upon the level of available reactive resources. It is therefore necessary to provide adequate reactive resources to support the desired level of power delivery.

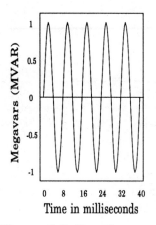

Figure 2-2 Reactive power.

2.4 TOTAL POWER

The instantaneous value of the total power is

$$s = P(1+\cos 2\omega t) + Q\sin 2\omega t)$$

The total power is the sum of the real power component and the reactive power component. Note that the total power flows away from the load; e.g., the total power is negative during certain periods of the waveform. This is due to the reactive power component which is negative during one-half of the cycle. The real power component represents the components of voltage and

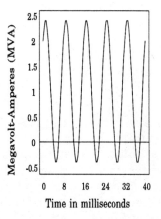

Figure 2-3 Total power.

current that are in phase, and the reactive power component represents the components of voltage and current that are in quadrature, e.g., that are 90° out of phase. The real and reactive power depend on the resistive, inductive, and capacitive nature of the circuit.

2.5 COMPLEX POWER (S)

The *complex power* S is by definition equal to the complex sum of the real and reactive power and is expressed as follows:

$$S = P + jQ \qquad (2\text{-}23)$$

The complex power may also be expressed as follows:

$$S = V\hat{I} \qquad (2\text{-}24)$$

The use of the conjugate current instead of the actual current is significant. The standard which has been adopted by the electric utility industry for the representation of complex power is

$$S = VI \angle \theta_V - \theta_I \qquad (2\text{-}25)$$

Note that the product of voltage and current would yield the following:

$$S = VI = VI \angle \theta_V + \theta_I \qquad (2\text{-}26)$$

It is therefore necessary to use the conjugate of the current phasor to achieve the desired result.

$$S = V\hat{I} = VI \angle \theta_V - \theta_I \qquad (2\text{-}27)$$

The complex power may also be expressed in terms of current and impedance as follows:

$$S = V\hat{I} = (IZ)\hat{I} = I\hat{I}Z \qquad (2\text{-}28)$$

Further reduction of the expression yields

$$\begin{aligned}I\hat{I}Z &= (I \angle \theta_I)(I \angle -\theta_I)(Z \angle \theta_Z) \\ &= I^2 Z \angle (\theta_I - \theta_I + \theta_Z) \\ &= I^2 Z \angle \theta_Z\end{aligned} \qquad (2\text{-}29)$$

Note that the product of the current phasor I and the conjugate of the current phasor $\hat{I}$ yields the square of the phasor magnitude I^2.

CHAPTER 2

2.6 POWER FACTOR

The power factor of an electric circuit is the ratio of the real power P in watts to the apparent power S in volt-amperes. The power factor is also equal to the cosine of the phase angle of the *power triangle* that is formed by the real power P which is the adjacent side, the reactive power Q which is the opposite side, and the apparent power S which is the hypotenuse. The pythagorean theorem may be used to express these relationships:

$$S=\sqrt{P^2+Q^2} \qquad \theta=\tan^{-1}\frac{Q}{P} \qquad PF=\cos\theta$$

Capacitive loads cause current to lead voltage and are referred to as leading power factor loads; conversely, inductive loads cause current to lag voltage and are referred to as lagging power factor loads.

2.6.1 Leading Power Factor

The voltage is taken as the reference in determining the leading or lagging condition. The leading condition describes the case in which current leads voltage as in a capacitive circuit. A voltage angle and a leading or positive current angle will yield a phase angle that will be negative. The apparent power vector will therefore lie in the fourth quadrant as illustrated in Fig. 2-5.

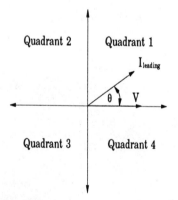

Figure 2-4 Voltage and current for leading power factor.

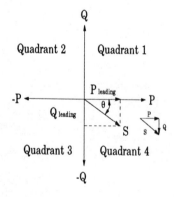

Figure 2-5 Real, reactive, and apparent power for leading power factor.

POWER

Consider that a transmission line is energized by the closure of a power circuit breaker at a local terminal. Electric charge must flow onto the open phase conductors in order to produce the difference in potential between the phase conductors and ground that is associated with the operating voltage of the system. The current which results from the flow of charge onto the open phase conductors is referred to as line charging. It is therefore evident that current must lead the voltage in order to charge the capacitance of the transmission line to the operating voltage of the system. This results in a leading power factor condition.

2.6.2 Lagging Power Factor

The lagging condition describes the case in which current lags voltage as in an inductive circuit. The voltage is again taken as the reference in determining the leading or lagging condition. A voltage angle and a lagging or negative current angle will yield a phase angle that will be positive. The apparent power vector will therefore lie in the first quadrant as illustrated in Fig. 2-7.

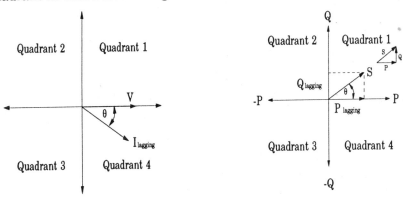

Figure 2-6 Voltage and current for lagging power factor.

Figure 2-7 Real, reactive and apparent power for lagging power factor.

Consider that the transmission line is placed under load by closure of the power circuit breaker at a remote terminal. This is referred to as paralleling the line since the line is placed in parallel with the electric

CHAPTER 2

power system. Loads on the electric power system are generally inductive which will cause the phase current to lag the applied voltage. This results in a lagging power factor condition.

2.7 ENERGY

The energy associated with the instantaneous power is given by the following expression:

$$E = \int s\,dt \qquad (2\text{-}30)$$

The integration of the instantaneous power over an interval of time yields the total energy provided to the electric circuit during that interval.

$$\begin{aligned}
E &= \int s\,dt \\
&= \int_{t1}^{t2} \{P[1+\cos(2\omega t)] + Q\sin(2\omega t)\}\,dt \\
&= P\int_{t1}^{t2} dt + \frac{P}{2\omega}\int_{t1}^{t2} \cos(2\omega t)(2\omega\,dt) + \frac{Q}{2\omega}\int_{t1}^{t2} \sin(2\omega t)(2\omega\,dt) \\
&= Pt\Big|_{t1}^{t2} + \frac{P}{2\omega}\sin(2\omega t)\Big|_{t1}^{t2} - \frac{Q}{2\omega}\cos(2\omega t)\Big|_{t1}^{t2}
\end{aligned} \qquad (2\text{-}31)$$

Note that the sine and cosine functions are circular in nature and that the integration of these functions over an integral number of cycles yields a value of zero. The energy associated with the instantaneous power for this condition therefore reduces to

$$E = Pt$$

where E is the energy in joules that is transferred during the interval, P is the average value of the real power component in joules per second, and t is the duration of the interval in seconds. Note that the reactive power component does not contribute to the energy that is dissipated in the load. The energy associated with the reactive power component is transferred between the electric fields that result from the application of the sinusoidal voltage between the phase conductors and ground and the magnetic fields that result from the flow of sinusoidal current through the phase conductors.

POWER

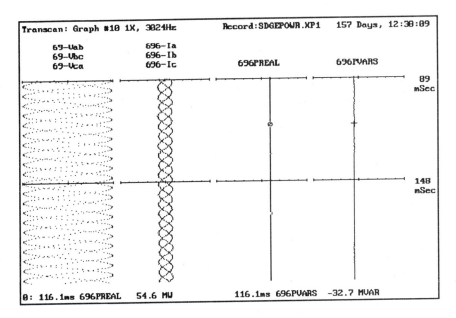

Figure 2-8 Digital fault recorder scan of real and reactive power. (*Courtesy of Mehta Tech, Inc.*)

2.8 MEASUREMENT OF REAL AND REACTIVE POWER

A digital recording of phase-to-phase voltages, phase currents, and real and reactive power is illustrated in Fig. 2-8. The Track and Mark section includes Marker 0 which shows the real power of 54.6 MW and the screen cursor which shows the reactive power of -32.7 MVAR. The phase angle associated with the megawatt and megavar values indicated in the Track and Mark section is computed as follows:

$$\theta = \tan^{-1}\frac{Q}{P} = \tan^{-1}\left(\frac{-32.7}{54.6}\right) = -31° \qquad (2\text{-}32)$$

The power factor is computed as follows:

$$PF = \cos(-31°) = 0.86 \qquad (2\text{-}33)$$

Engineers and technicians can check the reasonability of data which

CHAPTER 2

have been measured from the power system during the course of in-service testing by observing the quantities recorded in the electrical substation. The following steps are typical to determine the apparent impedance at the terminals of a transmission line:

1. The magnitude of the apparent impedance can be determined from the panel meters for the terminal voltages and phase currents as follows

$$Z_{apparent} = \frac{V}{I} \qquad (2\text{-}34)$$

2. The angle of the apparent impedance can be determined from the panel meters for real and reactive power as follows

$$\theta = \tan^{-1}\frac{Q}{P} \qquad (2\text{-}35)$$

CHAPTER 3

CIRCUIT ELEMENTS

3.1 RESISTANCE (R)

The resistance of a material is based on the electrical resistivity ρ which may be measured in ohm-inches (Ω-in). The resistance of a power system conductor is directly proportional to the resistivity and length of the conductor and inversely proportional to the cross-sectional area.

$$R = \rho \frac{l}{A} \qquad (3\text{-}1)$$

Consider an aluminum conductor which has a length of 100 mi and a cross-sectional area of 3 in^2. The total resistance in ohms[4] for the conductor would be:

$$\begin{aligned} R &= \rho \frac{l}{A} \\ &= \frac{(1.13 \times 10^{-6}\ \Omega\text{-in})(100\ \text{mi})(6.34 \times 10^6\ \text{in/mi})}{(3\ \text{in}^2)} \\ &= 2.4\ \Omega \end{aligned} \qquad (3\text{-}2)$$

Aluminum is often used for overhead transmission lines where weight is a consideration.

---

[4]The unit ohm (Ω) as a measure of resistance is named in honor of the German physicist Georg Simon Ohm (1787–1854) who formulated Ohm's law which states that the intensity of a constant electric current in a circuit is directly proportional to the electromotive force and inversely proportional to the resistance.

The total resistance for a copper conductor with the same dimensions would be

$$R = \rho \frac{l}{A}$$
$$= \frac{(6.63 \times 10^{-7} \ \Omega\text{-in})(100 \ \text{mi})(6.34 \times 10^6 \ \text{in/mi})}{(3 \ \text{in}^2)} \quad (3\text{-}3)$$
$$= 1.4 \ \Omega$$

The reciprocal of electrical resistivity is electrical conductivity σ which is measured in siemens (S).

3.1.1 Voltage and Current Relationships

The instantaneous voltage across the resistance is computed as the product of the instantaneous current and the resistance as follows:

$$v = iR \quad (3\text{-}4)$$

Let us assume a cosinusoidal current through the resistance of the following form:

$$i = I_m \cos(\omega t + \theta_I) \quad (3\text{-}5)$$

The voltage across the resistance is then computed as

$$v = iR = I_m R \cos(\omega t + \theta_I) \quad (3\text{-}6)$$

The maximum value of the voltage can be determined by inspection:

$$V_m = I_m R \quad (3\text{-}7)$$

The angle of the voltage can also be determined by inspection:

$$\theta_V = \theta_I \quad (3\text{-}8)$$

Since the voltage angle θ_V is equal to the current angle θ_I, then

$$\theta = \theta_V - \theta_I = 0 \quad (3\text{-}9)$$

A cosinusoidal current of the form $I_m \cos\omega t$ through the resistance R yields a cosinusoidal voltage of the form $V_m \cos\omega t$. There is therefore no phase shift between voltage and current in a purely resistive circuit. The resistance of power system conductors accounts for the real power losses in the system. The efficiency of the electric power system is therefore improved by the transmission of power at higher voltages which permits lower currents for the same power transfer and less real power losses in the resistance of the phase conductors.

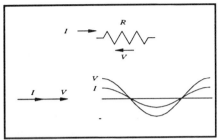

Figure 3-1 Resistance.

3.1.2 Power in Resistance

The instantaneous power through the resistance of the conductor is computed as the product of the instantaneous current and the instantaneous voltage and is expressed as follows:

$$p = vi \tag{3-10}$$

The instantaneous power may be expressed as a function of current and resistance as follows:

$$p = vi = (iR)i = i^2 R \tag{3-11}$$

Substituting the cosinusoidal representation for current yields

$$p = (I_m \cos\omega t)^2 R = I_m^2 R \cos^2 \omega t \tag{3-12}$$

The squared cosinusoid may be represented by the following trigonometric identity:

$$\cos^2 u = \frac{1}{2}(1 + \cos 2u) \tag{3-13}$$

The power delivered to the resistor may be expressed as follows:

$$p = \frac{I_m^2}{2} R (1 + \cos 2\omega t) \tag{3-14}$$

The current term may be expressed as follows:

$$\frac{I_m^2}{2} = \left(\frac{I_m}{\sqrt{2}}\right)^2 = I_{rms}^2 = I^2 \tag{3-15}$$

CHAPTER 3

The power to the resistor may therefore be expressed as follows:

$$p = I^2 R(1+\cos 2\omega t) \qquad (3\text{-}16)$$

Note that this is consistent with the real power component of the total power cited in Chap. 2 which is restated for convenience as follows:

$$\begin{aligned}
s &= VI\cos\theta(1+\cos 2\omega t) + VI\sin\theta\sin 2\omega t \\
  &= VI\cos(0)(1+\cos 2\omega t) + VI\sin(0)\sin 2\omega t \\
  &= VI(1+\cos 2\omega t) \\
  &= (IR)I(1+\cos 2\omega t) \\
  &= I^2 R(1+\cos 2\omega t)
\end{aligned} \qquad (3\text{-}17)$$

The average power to the resistor is therefore

$$P = I^2 R \qquad (3\text{-}18)$$

The instantaneous power may also be expressed as a function of voltage and resistance as follows:

$$p = vi = v\left(\frac{v}{R}\right) = \frac{v^2}{R} \qquad (3\text{-}19)$$

Substituting the cosinusoidal representation for voltage yields

$$p = \frac{(V_m \cos\omega t)^2}{R} = \frac{V_m^2 \cos^2\omega t}{R} \qquad (3\text{-}20)$$

The squared sinusoid may again be represented by the following trigonometric identity:

$$\cos^2 u = \frac{1}{2}(1+\cos 2u) \qquad (3\text{-}21)$$

The power delivered to the resistor may be expressed as follows:

$$p = \frac{V_m^2}{2R}(1+\cos 2\omega t) \qquad (3\text{-}22)$$

The voltage term may be expressed as follows:

$$\frac{V_m^2}{2} = \left(\frac{V_m}{\sqrt{2}}\right)^2 = V_{rms}^2 = V^2 \qquad (3\text{-}23)$$

CIRCUIT ELEMENTS 49

The power to the resistor may therefore be expressed as follows:

$$p = \frac{V^2}{R}(1+\cos 2\omega t) \qquad (3\text{-}24)$$

Note that this is also consistent with the real power component of the total power found in Chap. 2 which is restated for convenience as follows:

$$\begin{aligned}
s &= VI\cos\theta(1+\cos 2\omega t) + VI\sin\theta\sin 2\omega t \\
&= VI\cos(0)(1+\cos 2\omega t) + VI\sin(0)\sin 2\omega t \\
&= VI(1+\cos 2\omega t) \\
&= V\left(\frac{V}{R}\right)(1+\cos 2\omega t) \\
&= \frac{V^2}{R}(1+\cos 2\omega t)
\end{aligned} \qquad (3\text{-}25)$$

The average power to the resistor is therefore

$$P = \frac{V^2}{R} \qquad (3\text{-}26)$$

The total power in a purely resistive circuit therefore consists entirely of real power. The complex power S is computed as follows:

$$\begin{aligned}
S = V\hat{I} &= VI\angle\theta \\
&= VI\angle 0° \\
&= VI(\cos 0° + j\sin 0°) \\
&= VI \\
&= P
\end{aligned} \qquad (3\text{-}27)$$

3.2 INDUCTANCE (L)
The voltage that is induced in a conductor is based on the time rate of change of magnetic flux linkages.

$$v = \frac{d\Psi}{dt} \qquad (3\text{-}28)$$

The instantaneous voltage across the inductance of a circuit is given by the following expression:

$$v = L\frac{di}{dt} \qquad (3\text{-}29)$$

Equating these expressions yields

$$v = \frac{d\Psi}{dt} = L\frac{di}{dt} \qquad (3\text{-}30)$$

Solving for the inductance L yields

$$L = \frac{d\Psi}{dt} \cdot \frac{dt}{di} = \frac{d\Psi}{di} \qquad (3\text{-}31)$$

The inductance of a circuit in henries[5] (H) may therefore be expressed as the number of lines of magnetic flux produced for each ampere of current flow through the circuit.

3.2.1 Voltage and Current Relationships for Inductance

We may restate the voltage across the inductance as follows:

$$v = L\frac{di}{dt} \qquad (3\text{-}32)$$

Let us assume a cosinusoidal current through the inductance of the following form:

$$i = I_m \cos(\omega t + \theta_I) \qquad (3\text{-}33)$$

The voltage across the inductance is then computed as

$$v = L\frac{di}{dt} = L d\frac{[I_m \cos\omega t + \theta_I]}{dt} = -\omega L I_m \sin(\omega t + \theta_I) \qquad (3\text{-}34)$$

---

[5]The unit henry (H) as a measure of inductance is named in honor of the American physicist Joseph Henry (1797–1878) who demonstrated that the variation of a current of 1 A/s induces an electromotive force of 1 V.

CIRCUIT ELEMENTS 51

The resultant voltage may be expressed as a cosine function by use of the trigonometric identity $\sin\theta = \cos(\theta - \pi/2)$ as follows:

$$v = -\omega L I_m \sin(\omega t + \theta_I) = -\omega L I_m \cos\left(\omega t + \theta_I - \frac{\pi}{2}\right) \quad (3\text{-}35)$$

The negative sign may be absorbed into the argument of the function by adding $\pm\pi$ rad.

$$v = \omega L I_m \cos\left(\omega t + \theta_I + \frac{\pi}{2}\right) \quad (3\text{-}36)$$

A cosinusoidal current of the form $I_m\cos(\omega t+\theta_I)$ through the inductance L yields a cosinusoidal voltage of the form $V_m\cos(\omega t+\theta_I+\pi/2)$. The voltage therefore leads the current by $\pi/2$ rad or 90°. The flow of current through the phase conductors of the electric power system produces a magnetic field. The inductance of a conductor is a measure of the number of lines of magnetic flux per ampere of current flow.

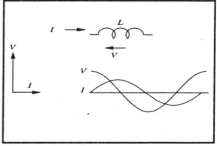

Figure 3-2 Inductance.

3.2.2 Inductive Reactance
We restate the voltage and current relationships as follows:

$$i = I_m \cos(\omega t + \theta_I) \quad (3\text{-}37)$$

$$v = \omega L I_m \cos(\omega t + \theta_I + \frac{\pi}{2}) \quad (3\text{-}38)$$

The voltage is of the form:

$$v = V_m \cos(\omega t + \theta_V) \quad (3\text{-}39)$$

The voltage magnitude may be stated by inspection:

$$V_m = \omega L I_m \quad (3\text{-}40)$$

52 CHAPTER 3

The voltage angle may also be stated by inspection:

$$\theta_V = \theta_I + \frac{\pi}{2} \tag{3-41}$$

The equations may be restated as follows:

$$V_m \cos(\omega t + \theta_V) = \omega L I_m \cos\left(\omega t + \theta_I + \frac{\pi}{2}\right) \tag{3-42}$$

Dividing both sides of the equation by $\sqrt{2}$ to obtain the rms magnitudes yields

$$\frac{V_m}{\sqrt{2}} \cos(\omega t + \theta_V) = \omega L \frac{I_m}{\sqrt{2}} \cos\left(\omega t + \theta_I + \frac{\pi}{2}\right) \tag{3-43}$$

$$V \cos(\omega t + \theta_V) = \omega L I \cos\left(\omega t + \theta_I + \frac{\pi}{2}\right) \tag{3-44}$$

The relationship between voltage and current may be expressed in exponential form as follows:

$$\Re V \varepsilon^{j(\omega t + \theta_V)} = \Re \omega L I \varepsilon^{j(\omega t + \pi/2)} \tag{3-45}$$

The nature of this identity requires that the imaginary parts of the functions are also equal. The reader is encouraged to examine this assertion by considering the following points:

1. The voltage is based on the real component of the voltage phasor which rotates at radian frequency ω and has an angular position of θ_V at time $t=0$.

2. The current is based on the real component of the current phasor which rotates at radian frequency ω and has an angular position of θ_I at time $t=0$.

3. The magnitude of the voltage phasor is related to the magnitude of the current phasor by the constant of proportionality ωL.

CIRCUIT ELEMENTS

4. The constant of proportionality is applied to both the real and imaginary components of the phasors.

The expression may therefore be written as follows:

$$V\varepsilon^{j(\omega t+\theta_v)} = \omega L I \varepsilon^{j(\omega t+\theta_i+\pi/2)} \qquad (3\text{-}46)$$

$$V = \omega L I \varepsilon^{j\pi/2} \qquad (3\text{-}47)$$

The remaining exponential term may be resolved as follows:

$$\begin{aligned}\varepsilon^{j\pi/2} &= \cos(\pi/2) + j\sin(\pi/2)\\ &= 0 + j1 \\ &= j\end{aligned} \qquad (3\text{-}48)$$

The relationship between voltage and current for the inductive circuit may be written as

$$V = j\omega L I \qquad (3\text{-}49)$$

The term ωL is known as the inductive reactance and is expressed as follows:

$$X_L = \omega L \qquad (3\text{-}50)$$

The relationship may therefore be restated as follows:

$$V = jX_L I \qquad (3\text{-}51)$$

3.2.3 Power in Inductance

The total power is restated as follows:

$$s = P(1+\cos 2\omega t) + Q\sin 2\omega t \qquad (3\text{-}52)$$

The current through the inductance lags the voltage by 90°. The phase angle is therefore

$$\theta = \theta_V - \theta_I = 0 - \left(-\frac{\pi}{2}\right) = \frac{\pi}{2} \qquad (3\text{-}53)$$

The real and reactive power are

$$P = VI\cos\theta = VI\cos\left(\frac{\pi}{2}\right) = 0 \qquad (3\text{-}54)$$

$$Q = VI\sin\theta = VI\sin\left(\frac{\pi}{2}\right) = VI \qquad (3\text{-}55)$$

The total power to the inductance is therefore

$$\begin{aligned}s &= P(1+\cos 2\omega t) + Q\sin 2\omega t \\ &= VI\sin 2\omega t\end{aligned} \qquad (3\text{-}56)$$

The total power in a purely inductive circuit therefore consists entirely of reactive power which is absorbed by the inductance of the circuit.

The complex power S is computed as follows:

$$\begin{aligned}S &= V\hat{I} = VI\angle\theta \\ &= VI(\cos 90° + j\sin 90°) \\ &= jVI\end{aligned} \qquad (3\text{-}57)$$

The complex power to the inductance is therefore

$$S = jQ \qquad (3\text{-}58)$$

Note that the conjugate of the current phasor in the complex power expression results in a positive reactive power. The reactive power is therefore taken to be positive for the lagging condition, e.g., the case in which current lags the applied voltage. The complex power may be expressed as a function of current and inductive reactance as follows:

$$S = jQ = jVI = j(IX_L)I = jI^2 X_L \qquad (3\text{-}59)$$

The complex power may also be expressed as a function of voltage and inductive reactance as follows:

$$S = jQ = jVI = jV\left(\frac{V}{X_L}\right) = j\frac{V^2}{X_L} \qquad (3\text{-}60)$$

1.2.4 Energy in Inductance

The energy in joules which is stored in the inductance of the circuit is computed as the summation of the product of the power in joules per second and the time in seconds over the interval of integration.

CIRCUIT ELEMENTS

The energy in inductance is expressed as follows:

$$W_L = \int vi\,dt \qquad (3\text{-}61)$$

Substituting the expressions for voltage and current in the inductive circuit yields

$$W_L = \int L\left(\frac{di}{dt}\right)(i)dt = \int Li\,di \qquad (3\text{-}62)$$

The solution of the integral yields

$$W_L = \frac{1}{2}Li^2 + W_{L0} \qquad (3\text{-}63)$$

where W_{L0} is the initial energy stored in the inductance of the circuit.

The energy stored in an inductor can also be derived from electromagnetic theory. The following is a brief discussion of the considerations involved in determining the energy which is stored in the static magnetic field that exists in the volume of the flux path of the inductor. The significant dimensions of the core of Fig. 3-3 are the mean length of the flux path l and the cross-sectional area A. These quantities are used to determine the flux density and the flux intensity in the core.

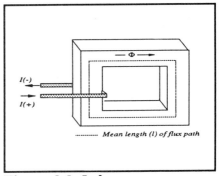

Figure 3-3 Inductor core.

The magnetic flux Φ in the core of the inductor may be expressed as follows:

$$\Phi = \oint B \cdot dA \qquad (3\text{-}64)$$

where B is the magnetic flux density in lines of flux per unit area and A is the cross-sectional area of the flux path.

CHAPTER 3

The solution of the integral for the uniform static magnetic field yields the following result:

$$\Phi = BA \tag{3-65}$$

The relationship may be rewritten to yield the flux density as follows:

$$B = \frac{\Phi}{A} \tag{3-66}$$

The total current enclosed by the magnetic field may be expressed as follows:

$$I = \oint H \cdot dl \tag{3-67}$$

where H is the magnetic field intensity and l is the mean length of the flux path.

The solution of the integral for the uniform static magnetic field yields the following result:

$$I = Hl \tag{3-68}$$

This relationship may be restated to express the magnetic flux intensity H in amperes[6] per unit length of the flux path as follows:

$$H = \frac{I}{l} \tag{3-69}$$

The magnetic flux density B is related to the magnetic flux intensity H by the magnetic permeability μ of the core material.

$$B = \mu H \tag{3-70}$$

---

[6]The unit ampere (A) is named in honor of the French physicist André Marie Amperè (1775-1836) who developed the science of electrodynamics.

CIRCUIT ELEMENTS

Substituting the parameters for B and H yields

$$\frac{\Phi}{A} = \mu \frac{I}{l} \tag{3-71}$$

Manipulating terms yields

$$\frac{\Phi}{I} = \mu \frac{A}{l} \tag{3-72}$$

Note that the number of lines of flux per ampere Φ/I is the measure of the inductance of the magnetic circuit. The inductance may therefore be stated in terms of the magnetic permeability μ, the cross-sectional area of the flux path A, and the mean length of the flux path l.

$$L = \mu \frac{A}{l} \tag{3-73}$$

The energy stored in the inductance may therefore be written as follows:

$$W_L = \frac{1}{2} L I^2 = \frac{1}{2} \left(\mu \frac{A}{l} \right) (H \cdot l)^2 = \frac{1}{2} \mu H^2 (A \cdot l) \tag{3-74}$$

The term $A \cdot l$ is the volume of the flux path. The energy which is stored in the inductance per unit volume w may therefore be expressed as

$$w_L = \frac{1}{2} \mu H^2 \tag{3-75}$$

3.3 CAPACITANCE (C)

The capacitance C may be measured by the amount of charge Q in coulombs[7] required to produce a potential difference of 1 V across the

---

[7] The unit coulomb (C) as a measure of charge is named in honor of the French physicist Charles Augustin de Coulomb (1736-1806) who quantified the amount of electrical charge provided by a current of 1 A flowing for 1 s.

CHAPTER 3

dielectric. The unit of capacitance is the farad (F).[8]

$$C = \frac{Q}{V} \tag{3-76}$$

The instantaneous current may be expressed as the time rate of change of electric charge as in the following expression:

$$i = \frac{dQ}{dt} \tag{3-77}$$

But

$$Q = CV \tag{3-78}$$

The expression for current can therefore be expressed as

$$i = \frac{dQ}{dt} = d\frac{(CV)}{dt} \tag{3-79}$$

Since C is a constant, the expression may be further developed as

$$i = C\frac{dV}{dt} \tag{3-80}$$

3.3.1 Voltage and Current Relationships

Let us assume a cosinusoidal voltage across the capacitance of the following form:

$$v = V_m \cos(\omega t + \theta_V) \tag{3-81}$$

The current through the capacitance is then computed as

$$i = C\frac{dv}{dt} = Cd\frac{[V_m \cos(\omega t + \theta_V)]}{dt} = -\omega C V_m \sin(\omega t + \theta_V) \tag{3-82}$$

---

[8]The unit farad (F) as a measure of capacitance is named in honor of the English physicist Michael Faraday (1791—1867) who discovered electromagnetic induction and formulated the laws of electrolysis.

CIRCUIT ELEMENTS

The resultant current may be expressed as a cosine function by use of the trigonometric identity sinu=cos(u-π/2) as follows:

$$i = -\omega C V_m \sin(\omega t + \theta_V) = -\omega C V_m \cos\left(\omega t + \theta_V - \frac{\pi}{2}\right) \quad (3\text{-}83)$$

The negative sign may be absorbed into the argument of the function by adding ±π rad.

$$i = \omega C V_m \cos\left(\omega t + \theta_V + \frac{\pi}{2}\right) \quad (3\text{-}84)$$

The magnitude of the current is determined by inspection as follows:

$$I_m = \omega C V_m \quad (3\text{-}85)$$

The angle of the current is also determined by inspection as follows:

$$\theta_I = \theta_V + \frac{\pi}{2} \quad (3\text{-}86)$$

A cosinusoidal voltage of the form $V_m(\cos\omega t + \theta_V)$ across the capacitance C yields a cosinusoidal current of the form $I_m(\cos\omega t + \theta_V + \pi/2)$. The current therefore leads the voltage by π/2 rad or 90°. The application of the operating voltage of the electric power system to the capacitance of the phase conductors produces an electric field. Reactive energy is stored in the electric field which is returned to the circuit when the source of voltage is removed.

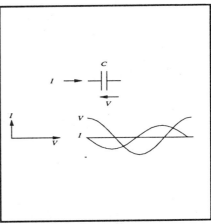

Figure 3-4 Capacitance.

3.3.2 Capacitive Reactance

We restate the voltage and current relationships as follows:

$$v = V_m \cos(\omega t + \theta_V) \tag{3-87}$$

$$i = \omega C V_m \cos\left(\omega t + \theta_V + \frac{\pi}{2}\right) \tag{3-88}$$

The current equation is of the form

$$i = I_m \cos(\omega t + \theta_I) \tag{3-89}$$

The magnitude of the current may be obtained by inspection:

$$I_m = \omega C V_m \tag{3-90}$$

The angle of the current may also be obtained by inspection:

$$\theta_I = \theta_V + \frac{\pi}{2} \tag{3-91}$$

The equations may be restated as follows:

$$V_m \cos(\omega t + \theta_V) = I_m \cos\left(\omega t + \theta_V + \frac{\pi}{2}\right) \tag{3-92}$$

Dividing both sides of the equation by $\sqrt{2}$ yields the rms magnitudes as follows:

$$\frac{I_m}{\sqrt{2}} \cos\left(\omega t + \theta_I + \frac{\pi}{2}\right) = \omega C \frac{V_m}{\sqrt{2}} \cos(\omega t + \theta_V) \tag{3-93}$$

$$I \cos\left(\omega t + \theta_I + \frac{\pi}{2}\right) = \omega C V \cos(\omega t + \theta_V) \tag{3-94}$$

The relationship between voltage and current may be expressed in exponential form as follows:

$$\Re I \varepsilon^{j(\omega t + \theta_I)} = \Re \omega C V \varepsilon^{j(\omega t + \theta_V + \pi/2)} \tag{3-95}$$

The nature of this identity requires that the imaginary parts of the functions are also equal for reasons cited previously.

CIRCUIT ELEMENTS

$$I\varepsilon^{j(\omega t+\theta_I)} = \omega CV\varepsilon^{j(\omega t+\theta_V+\pi/2)} \qquad (3\text{-}96)$$

$$I = \omega CV\varepsilon^{j\pi/2} \qquad (3\text{-}97)$$

The remaining exponential term may be resolved as follows:

$$\begin{aligned}\varepsilon^{j\pi/2} &= \cos\left(\frac{\pi}{2}\right) + j\sin\left(\frac{\pi}{2}\right) \\ &= 0 + j1 \\ &= j\end{aligned} \qquad (3\text{-}98)$$

The relationship between voltage and current for the capacitive circuit may be written as follows:

$$I = j\omega CV \qquad (3\text{-}99)$$

The relationship may be alternatively stated as follows:

$$V = \left(\frac{1}{j\omega C}\right)I \qquad (3\text{-}100)$$

The term $1/j\omega C$ is known as the capacitive reactance and is expressed as follows:

$$-jX_C = \frac{1}{j\omega C} \qquad (3\text{-}101)$$

The relationship may therefore be restated as follows:

$$V = -jX_C I \qquad (3\text{-}102)$$

3.3.3 Power in Capacitance

The total power expression is restated as follows:

$$p = P(1 + \cos 2\omega t) + Q\sin 2\omega t \qquad (3\text{-}103)$$

The current through the capacitance leads the voltage by 90°. The phase angle is therefore

$$\theta = \theta_V - \theta_I = 0 - \left(\frac{\pi}{2}\right) = -\frac{\pi}{2} \qquad (3\text{-}104)$$

CHAPTER 3

The real and reactive powers are

$$P = VI\cos\theta = VI\cos\left(-\frac{\pi}{2}\right) = 0 \qquad (3\text{-}105)$$

$$Q = VI\sin\theta = VI\sin\left(-\frac{\pi}{2}\right) = -VI \qquad (3\text{-}106)$$

The total power to the capacitance is therefore

$$\begin{aligned} p &= P(1+\cos 2\omega t) + Q\sin 2\omega t \\ &= -VI\sin 2\omega t \end{aligned} \qquad (3\text{-}107)$$

The total power in a purely capacitive circuit therefore consists entirely of reactive power which is produced by the capacitance of the circuit.

The complex power S is computed as follows:

$$\begin{aligned} S = V\hat{I} &= VI\angle\theta \\ &= VI[\cos(-90°) + j\sin(-90°)] \\ &= -jVI \end{aligned} \qquad (3\text{-}108)$$

The complex power to the capacitance is

$$S = -jQ \qquad (3\text{-}109)$$

Note that the conjugate of the current phasor in the complex power expression yields a negative reactive power. The reactive power is therefore taken to be negative for the leading condition.

The complex power may be expressed as a function of current and capacitive reactance as follows:

$$S = -jQ = -jVI = -j(IX_C)I = -jI^2X_C \qquad (3\text{-}110)$$

The complex power may also be expressed as a function of voltage and capacitive reactance as follows:

$$S = -jQ = -jVI = -jV\left(\frac{V}{X_C}\right) = -j\frac{V^2}{X_C} \qquad (3\text{-}111)$$

3.3.4 Energy in Capacitance

The energy in joules[9] which is stored in the capacitance of the circuit is computed as the summation of the product of the power in joules per second and the time in seconds and is expressed as follows:

$$W_C = \int vi\,dt = \int vC\left(\frac{dv}{dt}\right)dt = \int Cv\,dv \qquad (3\text{-}112)$$

The solution of the integral yields

$$W_C = \frac{1}{2}Cv^2 + W_{C0} \qquad (3\text{-}113)$$

where W_{C0} is the initial energy stored in the capacitance of the circuit.

The energy stored in a capacitor can also be derived from electromagnetic theory. The following is a brief discussion of the considerations involved in determining the energy which is stored in a static electric field that exists in the volume of the dielectric medium of the capacitor.

The electric flux Φ in the dielectric of the capacitor may be expressed as follows:

$$\Phi = \oint D \cdot dA \qquad (3\text{-}114)$$

where D is the electric flux density in lines of flux per unit area and A is the cross-sectional area of the flux path. The solution of the integral for the uniform static electric field yields the following result:

$$\Phi = DA \qquad (3\text{-}115)$$

The relationship may be rewritten to yield the flux density as follows:

---

[9]The unit joule (J) as a measure of energy is named in honor of the English physicist James Prescott Joule (1818–1889) who quantified the energy expended in 1 s by a current of 1 A at a potential difference of 1 V.

CHAPTER 3

$$D = \frac{\Phi}{A} \qquad (3\text{-}116)$$

The voltage V across the electric field may be expressed as follows:

$$V = \oint E \cdot dl \qquad (3\text{-}117)$$

where E is the electric field intensity and l is the mean length of the flux path. The solution of the integral for the uniform static electric field yields the following result:

$$V = El \qquad (3\text{-}118)$$

This relationship may be restated to express the electric flux intensity E in amperes per unit length of the flux path as follows:

$$E = \frac{V}{l} \qquad (3\text{-}119)$$

The electric flux density D is related to the electric flux intensity E by the electric permittivity ε of the dielectric material.

$$D = \varepsilon E \qquad (3\text{-}120)$$

Substituting the parameters for D and E yields

$$\frac{\Phi}{A} = \varepsilon \frac{V}{l} \qquad (3\text{-}121)$$

Manipulating terms yields

$$\frac{\Phi}{V} = \varepsilon \frac{A}{l} \qquad (3\text{-}122)$$

We will now invoke Gauss' law which states that the normal component of the electric flux density D over any closed surface equals the charge Q enclosed.

$$\Phi = Q \qquad (3\text{-}123)$$

The expression may be restated as follows:

$$\frac{Q}{V} = \varepsilon \frac{A}{l} \qquad (3\text{-}124)$$

CIRCUIT ELEMENTS 65

Note that the electric charge per volt Q/V is the measure of the capacitance of the electric circuit. The capacitance may therefore be stated in terms of the electric permittivity, the cross-sectional area of the flux path A, and the mean length of the flux path l.

$$C = \varepsilon \frac{A}{l} \qquad (3\text{-}125)$$

The energy stored in the capacitance may therefore be written as follows:

$$W_C = \frac{1}{2}CV^2 = \frac{1}{2}\left(\varepsilon\frac{A}{l}\right)(E \cdot l)^2 = \frac{1}{2}\varepsilon E^2(A \cdot l) \qquad (3\text{-}126)$$

The term $A \cdot l$ is the volume of the flux path. The energy which is stored in the capacitance per unit volume w may therefore be expressed as

$$w_C = \frac{1}{2}\varepsilon E^2 \qquad (3\text{-}127)$$

3.4 IMPEDANCE AND ADMITTANCE

We will establish the following conventions that will be used throughout the text to solve power system problems.

The impedance Z of a circuit may be expressed as follows:

$$Z = R + jX \qquad (3\text{-}128)$$

where $jX = jX_L - jX_C$. The reciprocal of impedance is the admittance Y which may be expressed as follows:

$$Y = G + jB \qquad (3\text{-}129)$$

These relationships for the circuit elements considered separately, e.g., as discrete components, are summarized in Table 3-1.

CHAPTER 3

Table 3-1 Impedance and Admittance Relationships

COMPONENT	IMPEDANCE (Z)	ADMITTANCE (Y)	$Z=1/Y$
Real	Resistance (R)	Conductance (G)	$R=1/G$
Imaginary	Reactance (X)	Susceptance (B)	$X=1/B$

The analysis of lumped parameters instead of discrete parameters requires the rationalization of the impedance to obtain the admittance as follows:

$$Y = \frac{1}{R+jX}$$
$$= \left(\frac{1}{R+jX}\right)\left(\frac{R-jX}{R-jX}\right) \qquad (3\text{-}130)$$
$$= \frac{R-jX}{R^2+X^2}$$

The admittance for the lumped parameter case may then be stated as

$$Y = G+jB$$
$$= \frac{R}{R^2+X^2} + j\frac{-X}{R^2+X^2} \qquad (3\text{-}131)$$

The conductance and susceptance may therefore be expressed as the real and imaginary components, respectively, of the admittance.

$$G = \frac{R}{R^2+X^2} \qquad (3\text{-}132)$$

$$B = -\frac{X}{R^2+X^2} \qquad (3\text{-}133)$$

These relationships will form the basis of the transmission line equations which will be developed and solved in Chap. 4.

CHAPTER 4

TRANSMISSION LINES

The transmission line parameters of resistance R, inductance L, capacitance C, and conductance G are distributed throughout the length of the line. The distributed model of a transmission line is illustrated in Fig. 4-1. This model is useful for determining the voltage and current conditions along the length of the line. The voltage at one terminal of a long transmission line, for example, may be within the operating limits, while the voltage at the other terminal may be above the high-voltage limit or below the low-voltage limit.

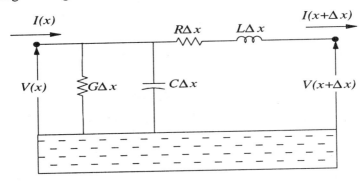

Figure 4-1 Distributed model.

The series impedance may be determined from the series resistance and the series inductive reactance.

$$Z = R + j\omega L \qquad (4\text{-}1)$$

The units of the series impedance for the distributed model are given in ohms per mile. The shunt admittance may be determined from the shunt conductance and the shunt susceptance.

CHAPTER 4

$$Y = G + j\omega C \quad (4\text{-}2)$$

The units of the shunt admittance are given in mhos per mile. The distributed model may be represented in terms of the series impedance and the shunt admittance as illustrated in Fig. 4-2.

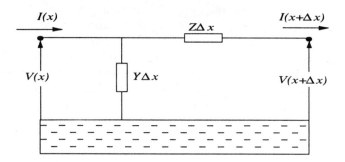

Figure 4-2 ZY model.

4.1 VOLTAGE AND CURRENT AS FUNCTIONS OF DISTANCE
4.1.1 Development of the Voltage Equation

The application of Kirchoff's voltage law to the sum of the voltages around the loop yields

$$V(x) - I(x + \Delta x) Z \Delta x - V(x + \Delta x) = 0 \quad (4\text{-}3)$$

Manipulating terms yields

$$\frac{V(x + \Delta x) - V(x)}{\Delta x} = -I(x + \Delta x) Z \quad (4\text{-}4)$$

Taking the limits of both sides of the equation as $\Delta x \to 0$,

$$\lim_{\Delta x \to 0} \frac{V(x + \Delta x) - V(x)}{\Delta x} = \lim_{\Delta x \to 0} -I(x + \Delta x) Z \quad (4\text{-}5)$$

The reader will recognize from elementary differential calculus that the left side of the equation may be resolved as follows:

$$\lim_{\Delta x \to 0} \frac{V(x + \Delta x) - V(x)}{\Delta x} = \frac{dV(x)}{dx} \quad (4\text{-}6)$$

TRANSMISSION LINES

The right side of the equation is resolved as follows:

$$\lim_{\Delta x \to 0} -I(x+\Delta x)Z = -I(x)Z \qquad (4\text{-}7)$$

The first derivative of voltage with respect to distance is therefore

$$\frac{dV(x)}{dx} = -I(x)Z \qquad (4\text{-}8)$$

We will assume from this point in the analysis that voltage and current are functions of distance. The first derivative of voltage with respect to distance is therefore stated as

$$\frac{dV}{dx} = -IZ \qquad (4\text{-}9)$$

The second derivative of voltage with respect to distance is obtained using the rules for differentiation of a product as follows:

$$\frac{d^2V}{dx^2} = -I\frac{dZ}{dx} - Z\frac{dI}{dx} \qquad (4\text{-}10)$$

Since the parameters Z and Y do not change along the length x of a uniform line, then

$$\frac{dZ}{dx} = \frac{dY}{dx} = 0 \qquad (4\text{-}11)$$

The second derivative of voltage therefore becomes

$$\frac{d^2V}{dx^2} = -Z\frac{dI}{dx} = (-Z)(-YV) = ZYV \qquad (4\text{-}12)$$

At this point we will introduce a new variable called the propagation constant γ which is expressed as follows:

$$\gamma = \sqrt{ZY} = \sqrt{(R+j\omega L)(G+j\omega C)} \qquad (4\text{-}13)$$

The second derivative of voltage can now be expressed as

$$\frac{d^2V}{dx^2} = \gamma^2 V \qquad (4\text{-}14)$$

4.1.2 Development of the Current Equation

The application of Kirchoff's current law to the sum of the node currents yields

$$I(x) - V(x)Y\Delta x - I(x+\Delta x) = 0 \qquad (4\text{-}15)$$

Manipulation of terms yields

$$\frac{I(x+\Delta x) - I(x)}{\Delta x} = -V(x)Y \qquad (4\text{-}16)$$

Taking the limit of both sides of the equation as $\Delta x \to 0$ yields

$$\lim_{\Delta x \to 0} \frac{I(x+\Delta x) - I(x)}{\Delta x} = \lim_{\Delta x \to 0} -V(x)Y \qquad (4\text{-}17)$$

The left side of the equation is resolved as follows:

$$\lim_{\Delta x \to 0} \frac{I(x+\Delta x) - I(x)}{\Delta x} = \frac{dI(x)}{dx} \qquad (4\text{-}18)$$

The term Δx does not appear in the right side of the equation, and it is therefore unaffected. The first derivative of current with respect to distance is therefore

$$\frac{dI(x)}{dx} = -V(x)Y \qquad (4\text{-}19)$$

We will again assume that the voltage and current are functions of distance. The first derivative of current with respect to distance is

$$\frac{dI}{dx} = -VY \qquad (4\text{-}20)$$

The second derivative of current with respect to distance is also obtained using the rules for differentiation of a product as follows:

$$\frac{d^2I}{dx^2} = -V\frac{dY}{dx} - Y\frac{dV}{dx} \qquad (4\text{-}21)$$

Again, since the parameters Z and Y do not change along the length x of a uniform line, then

$$\frac{dZ}{dx} = \frac{dY}{dx} = 0 \qquad (4\text{-}22)$$

TRANSMISSION LINES

The second derivative of current therefore becomes

$$\frac{d^2I}{dx^2} = -Y\frac{dV}{dx} = (-Y)(-ZI) = ZYI \qquad (4\text{-}23)$$

The second derivative of current can now be expressed as follows with the substitution of the propagation constant:

$$\frac{d^2I}{dx^2} = \gamma^2 I \qquad (4\text{-}24)$$

4.1.3 Solution of the Transmission Line Equations

The transmission line equations which were developed in the previous section may be solved using a variety of techniques. We will illustrate the solution by the method of linear operators.

4.1.3.1 General Solution. The transmission line equations for voltage and current as functions of distance are categorized as second-order, linear, homogeneous differential equations with constant coefficients. The general form of this type of equation is as follows:

$$a_0 D^2 y + a_1 Dy + a_2 y = 0 \qquad (4\text{-}25)$$

This equation is second-order because the highest-order derivative is the second derivative; linear because the power of y and its derivatives is of first degree; homogeneous because the right side of the equation is equal to zero; and the coefficients are constants because they are not functions of y. Note that the coefficients of the transmission line equations, e.g., the square of the propagation constant, are not functions of V, I, or x.

The voltage equation is restated as follows:

$$\frac{d^2V}{dx^2} = \gamma^2 V \qquad (4\text{-}26)$$

$$\frac{d^2V}{dx^2} - \gamma^2 V = 0 \qquad (4\text{-}27)$$

The auxiliary equation of the form $\phi(m)=0$ is

$$m^2-\gamma^2=0$$
$$(m+\gamma)(m-\gamma)=0 \qquad (4\text{-}28)$$
$$m=\pm\gamma$$

The general solution is then

$$V=V_1\varepsilon^{-\gamma x}+V_2\varepsilon^{\gamma x} \qquad (4\text{-}29)$$

where V_1 and V_2 are constants to be determined from the initial conditions. The solution of the current equation is found in the same manner. The solution is stated as follows:

$$\frac{d^2I}{dx^2}=\gamma^2 I \qquad (4\text{-}30)$$

$$\frac{d^2I}{dx^2}-\gamma^2 I=0 \qquad (4\text{-}31)$$

The auxiliary equation of the form $\phi(m)=0$ is

$$m^2-\gamma^2=0$$
$$(m+\gamma)(m-\gamma)=0 \qquad (4\text{-}32)$$
$$m=\pm\gamma$$

The general solution is then

$$I=I_1\varepsilon^{-\gamma x}+I_2\varepsilon^{\gamma x} \qquad (4\text{-}33)$$

where I_1 and I_2 are again constants to be determined by the initial conditions.

4.1.3.2 Particular Solution. The determination of specific values for the constants of the general solution yields the particular solution. The values of these constants must be determined by the initial conditions. At this point a discussion of initial conditions is appropriate. The principal equations are restated for convenience as follows:

$$V=V_1\varepsilon^{-\gamma x}+V_2\varepsilon^{\gamma x} \qquad (4\text{-}34)$$

$$I=I_1\varepsilon^{-\gamma x}+I_2\varepsilon^{\gamma x} \qquad (4\text{-}35)$$

TRANSMISSION LINES

The first term of the voltage solution $V_1\varepsilon^{-\gamma x}$ corresponds to the first term of the current solution $I_1\varepsilon^{-\gamma x}$ in the above equations. These traveling waves propagate in the positive x direction.
Similarly, the second term of the voltage solution $V_2\varepsilon^{\gamma x}$ corresponds to the second term of the current solution $I_2\varepsilon^{\gamma x}$ in the above equations. These traveling waves propagate in the negative x direction.

We now solve for I_1. The first derivative of the voltage equation is restated for convenience:

$$\frac{dV}{dx} = -IZ \tag{4-36}$$

Substituting $V_1\varepsilon^{-\gamma x}$ for V and $I_1\varepsilon^{-\gamma x}$ for I yields

$$d\frac{V_1\varepsilon^{-\gamma x}}{dx} = -I_1\varepsilon^{-\gamma x}Z \tag{4-37}$$

Taking the derivative of the voltage yields

$$-\gamma V_1\varepsilon^{-\gamma x} = -I_1\varepsilon^{-\gamma x}Z \tag{4-38}$$

Dividing both sides of the equation by $\varepsilon^{-\gamma x}$ and solving for I_1:

$$I_1 = \gamma\frac{V_1}{Z} \tag{4-39}$$

We now introduce a new variable called the characteristic impedance Z_C which represents the ratio of voltage to current at a given point on the transmission line.

$$Z_C = \sqrt{\frac{Z}{Y}} = \sqrt{\frac{R+j\omega L}{G+j\omega C}} \tag{4-40}$$

Since $\gamma=\sqrt{ZY}$ and $Z_C=\sqrt{Z/Y}$, then

$$I_1 = \gamma\frac{V_1}{Z} = \sqrt{ZY}\frac{V_1}{Z} = \frac{\sqrt{Z}\sqrt{Y}}{\sqrt{Z}\sqrt{Z}}V_1 = \frac{\sqrt{Y}}{\sqrt{Z}}V_1 = \frac{1}{\sqrt{Z/Y}}V_1 = \frac{V_1}{Z_C} \tag{4-41}$$

We solve for I_2 in a similar manner by substituting $V_2\varepsilon^{\gamma x}$ for V and $I_2\varepsilon^{\gamma x}$ for I.

CHAPTER 4

$$\frac{dV}{dx} = \frac{dV_2 \varepsilon^{\gamma x}}{dx} = \gamma V_2 \varepsilon^{\gamma x} = -I_2 \varepsilon^{\gamma x} Z \qquad (4\text{-}42)$$

$$I_2 = -\frac{V_2}{Z_C} \qquad (4\text{-}43)$$

The general solutions for the voltage and current equations become

$$V = V_1 \varepsilon^{-\gamma x} + V_2 \varepsilon^{\gamma x} \qquad (4\text{-}44)$$

$$I = \frac{V_1}{Z_C} \varepsilon^{-\gamma x} - \frac{V_2}{Z_C} \varepsilon^{\gamma x} \qquad (4\text{-}45)$$

The particular solutions for the voltage and current equations when $x=0$ are

$$V_0 = V(0) = V_1 \varepsilon^{-\gamma(0)} + V_2 \varepsilon^{\gamma(0)} = V_1 + V_2 \qquad (4\text{-}46)$$

$$I_0 = I(0) = \frac{V_1}{Z_C} \varepsilon^{-\gamma(0)} - \frac{V_2}{Z_C} \varepsilon^{\gamma(0)} = \frac{V_1}{Z_C} - \frac{V_2}{Z_C} \qquad (4\text{-}47)$$

Solving simultaneous equations for V_1 and V_2 yields

$$V_1 = \frac{1}{2}(V_0 + Z_C I_0) \qquad (4\text{-}48)$$

$$V_2 = \frac{1}{2}(V_0 - Z_C I_0) \qquad (4\text{-}49)$$

Substituting the above expressions for V_1 and V_2 into the general solutions to obtain the particular solutions at $x=0$:

$$V(x) = \frac{1}{2}(V_0 + Z_C I_0)\varepsilon^{-\gamma x} + \frac{1}{2}(V_0 - Z_C I_0)\varepsilon^{\gamma x} \qquad (4\text{-}50)$$

$$I(x) = \frac{1}{2Z_C}(V_0 + Z_C I_0)\varepsilon^{-\gamma x} - \frac{1}{2Z_C}(V_0 - Z_C I_0)\varepsilon^{\gamma x} \qquad (4\text{-}51)$$

TRANSMISSION LINES

These expressions may be rewritten as follows:

$$V(x) = V_0\left[\frac{1}{2}(\varepsilon^{-\gamma x} + \varepsilon^{\gamma x})\right] + Z_C I_0\left[\frac{1}{2}(\varepsilon^{-\gamma x} - \varepsilon^{\gamma x})\right] \quad (4\text{-}52)$$

$$I(x) = I_0\left[\frac{1}{2}(\varepsilon^{-\gamma x} + \varepsilon^{\gamma x})\right] + \frac{V_0}{Z_C}\left[\frac{1}{2}(\varepsilon^{-\gamma x} - \varepsilon^{\gamma x})\right] \quad (4\text{-}53)$$

The following identities may be invoked for this expression:

$$\cosh u = \tfrac{1}{2}(\varepsilon^u + \varepsilon^{-u}) \quad (4\text{-}54)$$

$$\sinh u = \tfrac{1}{2}(\varepsilon^u - \varepsilon^{-u}) \quad (4\text{-}55)$$

The voltage and current equations may be rewritten as follows:

$$V(x) = V_0 \cosh(\gamma x) - Z_C I_0 \sinh(\gamma x) \quad (4\text{-}56)$$

$$I(x) = I_0 \cosh(\gamma x) - \frac{V_0}{Z_C} \sinh(\gamma x) \quad (4\text{-}57)$$

4.2 VOLTAGE AND CURRENT AS FUNCTIONS OF DISTANCE AND TIME

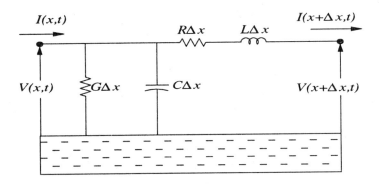

Figure 4-3 Distributed model $V(x, t)$, $I(x, t)$

CHAPTER 4

4.2.1 Development of the Voltage Equation

The application of Kirchoff's voltage law to the sum of the voltages around the loop yields

$$V(x,t) - R\Delta x I(x+\Delta x,t) - L\Delta x \frac{\partial I(x+\Delta x,t)}{\partial t} - V(x+\Delta x,t) = 0 \quad (4\text{-}58)$$

Manipulation of the terms yields

$$\frac{V(x+\Delta x,t) - V(x,t)}{\Delta x} = -RI(x+\Delta x,t) - L\frac{\partial I(x+\Delta x,t)}{\partial t} \quad (4\text{-}59)$$

Again from elementary differential calculus:

$$\lim_{\Delta x \to 0} \frac{V(x+\Delta x,t) - V(x,t)}{\Delta x} = -\frac{\partial V(x,t)}{\partial x} \quad (4\text{-}60)$$

$$\lim_{\Delta x \to 0} -RI(x+\Delta x,t) - L\frac{\partial I(x+\Delta x,t)}{\partial t} = -RI(x,t) - L\frac{\partial I(x,t)}{\partial t} \quad (4\text{-}61)$$

The first derivative of voltage with respect to distance is therefore

$$\frac{\partial V(x,t)}{\partial x} = -RI(x,t) - L\frac{\partial I(x,t)}{\partial t} \quad (4\text{-}62)$$

4.2.2 Development of the Current Equation

The application of Kirchoff's current law to the sum of the currents into the node yields

$$I(x,t) - G\Delta x V(x,t) - C\Delta x \frac{\partial V(x,t)}{\partial t} - I(x+\Delta x,t) = 0 \quad (4\text{-}63)$$

Manipulation of the terms yields

$$\frac{I(x+\Delta x,t) - I(x,t)}{\Delta x} = -GV(x,t) - C\frac{\partial V(x,t)}{\partial t} \quad (4\text{-}64)$$

From elementary differential calculus:

$$\lim_{\Delta x \to 0} \frac{I(x+\Delta x,t) - I(x,t)}{\Delta x} = \frac{\partial I(x,t)}{\partial t} \quad (4\text{-}65)$$

This yields the first derivative of current with respect to time.

TRANSMISSION LINES

$$\frac{\partial I(x,t)}{\partial x} = -GV(x,t) - C\frac{\partial V(x,t)}{\partial t} \qquad (4\text{-}66)$$

4.2.3 Substitution of Derivatives
The first partial derivatives of voltage and current were found to be

$$\frac{\partial V(x,t)}{\partial x} = -RI(x,t) - L\frac{\partial I(x,t)}{\partial t} \qquad (4\text{-}67)$$

$$\frac{\partial I(x,t)}{\partial x} = -GV(x,t) - C\frac{\partial V(x,t)}{\partial t} \qquad (4\text{-}68)$$

The second partial derivatives of voltage and current are

$$\frac{\partial^2 V(x,t)}{\partial x^2} = -R\frac{\partial I(x,t)}{\partial x} - L\frac{\partial^2 I(x,t)}{\partial x \partial t} \qquad (4\text{-}69)$$

$$\frac{\partial^2 I(x,t)}{\partial x^2} = -G\frac{\partial V(x,t)}{\partial x} - C\frac{\partial^2 V(x,t)}{\partial x \partial t} \qquad (4\text{-}70)$$

We will first proceed to rewrite the voltage equation with substitution of terms. The first term on the right side of the voltage equation can be expressed as

$$R\frac{\partial I(x,t)}{\partial x} = R\left[GV(x,t) + C\frac{\partial V(x,t)}{\partial t}\right] \qquad (4\text{-}71)$$

The second term on the right side of the equation can be expressed as

$$L\frac{\partial^2 I(x,t)}{\partial x \partial t} = L\left\{\frac{\partial\left[\frac{\partial I(x,t)}{\partial x}\right]}{\partial t}\right\}$$

$$= L\left\{\frac{\partial\left[GV(x,t) + C\frac{\partial V(x,t)}{\partial t}\right]}{\partial t}\right\} \qquad (4\text{-}72)$$

$$= L\left[G\frac{\partial V(x,t)}{\partial t} + C\frac{\partial^2 V(x,t)}{\partial t^2}\right]$$

The voltage equation can therefore be expressed as follows:

$$\frac{\partial^2 V(x,t)}{\partial x^2} = RGV(x,t) + RC\frac{\partial V(x,t)}{\partial t} + LG\frac{\partial V(x,t)}{\partial t} + LC\frac{\partial^2 V(x,t)}{\partial t^2}$$
$$= RGV(x,t) + [RC+LG]\frac{\partial V(x,t)}{\partial t} + LC\frac{\partial^2 V(x,t)}{\partial t^2} \quad (4\text{-}73)$$

We will now rewrite the current equation with substitution of terms. The first term on the right side of the current equation can be expressed as

$$G\frac{\partial V(x,t)}{\partial x} = G\left[RI(x,t) + L\frac{\partial I(x,t)}{\partial t}\right] \quad (4\text{-}74)$$

The second term on the right side of the equation can be expressed as

$$C\frac{\partial^2 V(x,t)}{\partial x \partial t} = C\left\{\frac{\partial\left[\frac{\partial V(x,t)}{\partial x}\right]}{\partial t}\right\}$$
$$= C\left\{\frac{\partial\left[RI(x,t) + L\frac{\partial I(x,t)}{\partial t}\right]}{\partial t}\right\} \quad (4\text{-}75)$$
$$= C\left[R\frac{\partial I(x,t)}{\partial t} + L\frac{\partial^2 I(x,t)}{\partial t^2}\right]$$

The current equation can therefore be expressed as follows:

$$\frac{\partial^2 I(x,t)}{\partial x^2} = RGI(x,t) + LG\frac{\partial I(x,t)}{\partial t} + RC\frac{\partial I(x,t)}{\partial t} + LC\frac{\partial^2 I(x,t)}{\partial t^2}$$
$$= RGI(x,t) + [LG+RC]\frac{\partial I(x,t)}{\partial t} + LC\frac{\partial^2 I(x,t)}{\partial t^2} \quad (4\text{-}76)$$

4.2.4 Solution of the Voltage and Current Equations as Functions of Distance and Time

The equations for voltage and current as functions of distance and time are referred to as the telegrapher's equations. They are restated for convenience.

TRANSMISSION LINES

$$\frac{\partial^2 V(x,t)}{\partial x^2} = RGV(x,t) + [RC+LG]\frac{\partial V(x,t)}{\partial t} + LC\frac{\partial^2 V(x,t)}{\partial t^2} \quad (4\text{-}77)$$

$$\frac{\partial^2 I(x,t)}{\partial x^2} = RGI(x,t) + [LG+RC]\frac{\partial I(x,t)}{\partial t} + LC\frac{\partial^2 I(x,t)}{\partial t^2} \quad (4\text{-}78)$$

The solution of these equations is greatly simplified by neglecting the series resistance R and the shunt conductance G which are generally orders of magnitude less than the series inductive reactance ωL and the shunt susceptance ωC.

The equations may then be restated as follows:

$$\frac{\partial^2 V(x,t)}{\partial x^2} = LC\frac{\partial^2 V(x,t)}{\partial t^2} \quad (4\text{-}79)$$

$$\frac{\partial^2 I(x,t)}{\partial x^2} = LC\frac{\partial^2 I(x,t)}{\partial t^2} \quad (4\text{-}80)$$

The classical solution method for this equation involves the following transformation of variables:

$$\partial^2 \frac{V(w,u)}{\partial w \partial u} = 0 \quad (4\text{-}81)$$

where $w = x + vt$ and $u = x - vt$. This must necessarily be true since each successive differentiation of the voltage with respect to one variable regards the other variable as a constant. The successive integration of this term with respect to w and with respect to u yields

$$\begin{aligned}V(x,t) &= V(w) + V(u) + V_0 \\ &= V(x+vt) + V(x-vt) + V_0\end{aligned} \quad (4\text{-}82)$$

where V_0 represents the dc component. The partial derivative of the voltage with respect to distance may be stated as follows:

$$\frac{\partial V(x,t)}{\partial x} = \frac{\partial V(w)}{\partial w} \cdot \frac{\partial w}{\partial x} + \frac{\partial V(u)}{\partial u} \cdot \frac{\partial u}{\partial x} \quad (4\text{-}83)$$

CHAPTER 4

Note that

$$\frac{\partial w}{\partial x} = \frac{\partial (x+vt)}{\partial x} = 1 \qquad (4\text{-}84)$$

$$\frac{\partial u}{\partial x} = \frac{\partial (x-vt)}{\partial x} = 1 \qquad (4\text{-}85)$$

The partial derivative of the voltage with respect to distance is then:

$$\frac{\partial V(x,t)}{\partial x} = V'(w) + V'(u) \qquad (4\text{-}86)$$

The second derivative of the voltage with respect to distance is

$$\frac{\partial^2 V(x,t)}{\partial x^2} = V''(w) + V''(u) \qquad (4\text{-}87)$$

The partial derivative of the voltage with respect to time may be stated as follows:

$$\frac{\partial V(x,t)}{\partial t} = \frac{\partial V(w)}{\partial w} \cdot \frac{\partial w}{\partial t} + \frac{\partial V(u)}{\partial u} \cdot \frac{\partial u}{\partial t} \qquad (4\text{-}88)$$

Note that

$$\frac{\partial w}{\partial t} = \frac{\partial (x+vt)}{\partial t} = v \qquad (4\text{-}89)$$

$$\frac{\partial u}{\partial t} = \frac{\partial (x-vt)}{\partial t} = -v \qquad (4\text{-}90)$$

The partial derivative of the voltage with respect to time is then

$$\frac{\partial V(x,t)}{\partial t} = vV'(w) - vV'(u) \qquad (4\text{-}91)$$

The second derivative of the voltage with respect to time is

$$\frac{\partial^2 V(x,t)}{\partial t^2} = v^2 V''(w) + v^2 V''(u) \qquad (4\text{-}92)$$

The following solution can be established by comparison of the partial derivative of voltage with respect to distance and the partial derivative of voltage with respect to time.

TRANSMISSION LINES

$$\frac{\partial^2 V(x,t)}{\partial x^2} = \frac{1}{v^2}\frac{\partial^2 V(x,t)}{\partial t^2} \qquad (4\text{-}93)$$

A comparison of this result with the original voltage equation yields

$$\frac{1}{v^2} = LC \qquad (4\text{-}94)$$

$$v = \frac{1}{\sqrt{LC}} \qquad (4\text{-}95)$$

An examination of $V(w)=V(0)$ yields

$$x - vt = 0 \qquad (4\text{-}96)$$

$$x = vt \qquad (4\text{-}97)$$

This is a voltage function which propagates in the positive x direction with a velocity v. An examination of $V(u)=V(0)$ yields

$$x + vt = 0 \qquad (4\text{-}98)$$

$$x = -vt \qquad (4\text{-}99)$$

This is a voltage function which propagates in the negative x direction with a velocity $-v$. The solution is identical for the current equations.

4.3 TERMINAL CONDITIONS FOR THE TRANSMISSION LINE EQUATIONS

The transmission line equations which were developed and solved in the previous sections may be applied to the analysis of practical transmission line problems. The voltage and current equations are restated as follows:

$$V(x) = V_0 \cosh(\gamma x) - I_0 \frac{Z}{\gamma}\sinh(\gamma x) \qquad (4\text{-}100)$$

$$I(x) = I_0 \cosh(\gamma x) - V_0 \frac{Y}{\gamma}\sinh(\gamma x) \qquad (4\text{-}101)$$

CHAPTER 4

Table 4-1 Receiving-End Voltage and Current

QUANTITY	DESCRIPTION	CONVENTION
$V(x)$	Receiving-end voltage	V_R
V_0	Sending-end voltage	V_S
I_0	Sendinging-end current	I_S
x	Line length	l

4.3.1 Receiving-End Voltage and Current

The transmission line equations may be written to express the receiving-end voltage and current in terms of the sending-end quantities. The sending-end bus is taken as the point of origin where $x=0$ and the receiving-end bus is taken to be the point of consideration where $x=l$. These conventions are summarized in Table 4-1.

The voltage and current equations may be restated with substitution of the quantities cited in the above table to represent conditions at the receiving end as follows:

$$V_R = V_S \cosh(\gamma l) - I_S \frac{Z}{\gamma} \sinh(\gamma l) \qquad (4\text{-}102)$$

$$I_R = I_S \cosh(\gamma l) - V_S \frac{Y}{\gamma} \sinh(\gamma l) \qquad (4\text{-}103)$$

These equations may be further resolved by the following manipulations:

$$\frac{Z}{\gamma} = \frac{\sqrt{Z}\sqrt{Z}}{\sqrt{Z}\sqrt{Y}} = \frac{\sqrt{Z}}{\sqrt{Y}} = \sqrt{\frac{Z}{Y}} = Z_C \qquad (4\text{-}104)$$

TRANSMISSION LINES

$$\frac{Y}{\gamma} = \frac{\sqrt{Y}\sqrt{Y}}{\sqrt{Z}\sqrt{Y}} = \frac{\sqrt{Y}}{\sqrt{Z}} = \sqrt{\frac{Y}{Z}} = \frac{1}{Z_C} \quad (4\text{-}105)$$

The transmission line equations may then be rewritten as follows:

$$V_R = V_S \cosh(\gamma l) - I_S Z_C \sinh(\gamma l) \quad (4\text{-}106)$$

$$I_R = I_S \cosh(\gamma l) - \frac{V_S}{Z_C} \sinh(\gamma l) \quad (4\text{-}107)$$

4.3.2 Sending-End Voltage and Current

The transmission line equations may also be written to express the sending-end voltage. The receiving-end bus is taken as the point of origin where $x=0$ and the sending-end bus is taken to be the point of consideration where $x=-l$. These conventions are summarized in Table 4-2.

The voltage and current may be restated to represent conditions at the sending end as follows:

$$V_S = V_R \cosh(-\gamma l) - I_R Z_C \sinh(-\gamma l) \quad (4\text{-}108)$$

$$I_S = I_R \cosh(\gamma l) - \frac{V_R}{Z_C} \sinh(-\gamma l) \quad (4\text{-}109)$$

Since $\cosh(\gamma l) = \cosh(-\gamma l)$ and $\sinh(-\gamma l) = -\sinh(\gamma l)$, then the sending-end

Table 4-2 Sending-End Voltage and Current

QUANTITY	DESCRIPTION	CONVENTION
$V(x)$	Sending-end voltage	V_S
V_0	Receiving-end voltage	V_R
I_0	Receiving-end current	I_R
x	Line length	$-l$

equations may be restated as follows:

$$V_S = V_R \cosh(\gamma l) + I_R Z_C \sinh(\gamma l) \qquad (4\text{-}110)$$

$$I_S = \frac{V_R}{Z_C} \sinh(\gamma l) + I_R \cosh(\gamma l) \qquad (4\text{-}111)$$

4.3.3 Transmission Line Constants (A, B, C, and D)

The voltage and current equations for the sending-end and receiving-end buses may be simplified by assigning the following

$$A = \cosh(\gamma l) \qquad B = Z_C \sinh(\gamma l)$$

$$C = \frac{1}{Z_C} \sinh(\gamma l) \qquad D = \cosh(\gamma l)$$

The sending-end equations may be rewritten in terms of A, B, C, and D constants as follows:

$$V_S = AV_R + BI_R \qquad (4\text{-}112)$$

$$I_S = CV_R + DI_R \qquad (4\text{-}113)$$

The receiving-end equations may also be rewritten in terms of A, B, C, and D constants as follows:

$$V_R = AV_S - BI_S \qquad (4\text{-}114)$$

$$I_R = -CV_S + DI_S \qquad (4\text{-}115)$$

4.4 EQUIVALENT TRANSMISSION LINE MODELS

We have already established that the distributed parameter model of a transmission line is useful in determining the voltage and current conditions along the length of the line. The *lumped-parameter* models combine the distributed inductance, resistance, capacitance, and conductance into discrete circuit elements and are suitable for analyses in which it is not necessary to study the conditions along the length of the line but only the conditions at the line terminals. The two types of lumped-parameter models that are commonly employed in the analysis of electric power systems are the *pi model* (Fig. 4-4) and the *T model* (Fig. 4-5). We will now examine the equivalent and approximate forms for these models.

4.4.1 Equivalent Pi Model and T Model

The equivalent pi model and T model yield the values of voltage and current at the terminals of the transmission line that are equivalent to those obtained with the distributed model. We will develop the *ABCD* constants for an equivalent pi model and will leave it to the reader to develop those for the equivalent T model.

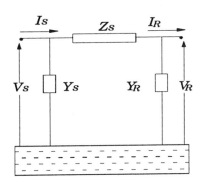

Figure 4-4 Equivalent pi model.

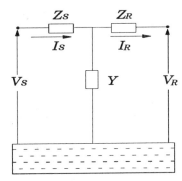

Figure 4-5 Equivalent T model.

The determination of the *ABCD* constants for the equivalent pi model is based on the application of Kirchoff's voltage and current laws to the nodes and branches of the model. The coefficients of the sending and receiving voltages and currents are then equated with those of the distributed model to determine the series impedance and the shunt admittances at the sending and receiving terminals. The receiving voltage may be expressed as follows:

$$V_R = V_S - (I_S - V_S Y_S)Z$$
$$= (1 + Y_S Z)V_S - Z I_S \qquad (4\text{-}116)$$

CHAPTER 4

This expression may be written in the following form:

$$V_R = AV_S - BI_S \qquad (4\text{-}117)$$

where $A=1+Y_S Z$ and $B=Z$. Equating the A and B coefficients with those of the distributed model yields

$$A = 1 + Y_S Z = \cosh\gamma l \qquad (4\text{-}118)$$

$$B = Z = Z_C \sinh\gamma l \qquad (4\text{-}119)$$

Solving for the sending-end admittance Y_S yields

$$Y_S = \frac{\cosh\gamma l - 1}{Z_C \sinh\gamma l} \qquad (4\text{-}120)$$

We will now invoke the following identity for hyperbolic functions:

$$\tanh\left(\frac{u}{2}\right) = \frac{\cosh u - 1}{\sinh u} \qquad (4\text{-}121)$$

The shunt admittance may then be expressed as follows:

$$Y_S = \frac{1}{Z_C}\tanh\frac{\gamma l}{2} \qquad (4\text{-}122)$$

The next step is to determine the receiving-end current as follows:

$$I_R = I_S - V_S Y_S - V_R Y_R \qquad (4\text{-}123)$$

Substituting the expression for V_R yields

$$\begin{aligned}I_R &= I_S - V_S Y_S - [V_S(1+Y_S Z) - I_S Z]Y_R \\ &= -(Y_S + Y_R + Y_S Y_R Z)V_S + (1+Y_R Z)I_S\end{aligned} \qquad (4\text{-}124)$$

This expression may be written in the following form:

$$I_R = -CV_S + DI_S \qquad (4\text{-}125)$$

Equating the C and D coefficients with the distributed model yields

$$C = Y_S + Y_R + Y_S Y_R Z = \frac{1}{Z_C}\sinh\gamma l \qquad (4\text{-}126)$$

$$D = 1 + Y_R Z = \cosh\gamma l \qquad (4\text{-}127)$$

We can now determine the receiving-end admittance as follows:

$$1+Y_S Z = \cosh\gamma l$$
$$1+Y_R Z = \cosh\gamma l \qquad (4\text{-}128)$$
$$Y_S = Y_R$$

The equivalent pi model with the associated series impedance and the shunt admittances at the sending and receiving terminals is illustrated in Fig. 4-6.

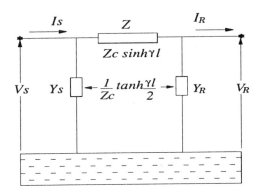

Figure 4-6 Equivalent pi model.

4.4.2 Approximate Pi Model and T Model

The approximate pi model (Fig. 4-7) is based on the following assumptions:

1. The shunt conductance is neglected.

2. The series impedance is represented by the lumped-parameter values of the series resistance and the series inductive reactance.

3. The lumped-parameter value of capacitance is equally divided and represented at the sending and receiving terminals. Note that the capacitive reactance at each terminal will therefore be double the value that would be obtained for the single

CHAPTER 4

lumped-parameter capacitance due to the rules regarding equal impedances in parallel.

The approximate T model (Fig. 4-8) is based on the following assumptions:

1. The shunt conductance is neglected.

2. The lumped-parameter value of the shunt capacitance is represented at the center of the transmission line.

3. The series resistance and inductive reactance are divided equally and connected between the sending and receiving terminals and the shunt capacitance.

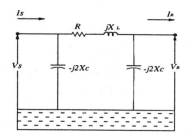
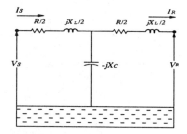

Figure 4-7 Approximate pi model.

Figure 4-8 Approximate T model.

The approximate models provide results which are not significantly different from those obtained with the exact models for most transmission lines with the advantage that they are easier to use.

4.5 CHARACTERISTIC IMPEDANCE

The ratio of voltage to current associated with a traveling wave along the transmission line is defined as the characteristic impedance Z_c and is computed as follows:

TRANSMISSION LINES

$$\frac{V}{I} = \frac{IZ}{VY}$$

$$\frac{V^2}{I^2} = \frac{Z}{Y}$$
(4-129)

$$Z_C = \frac{V}{I} = \sqrt{\frac{V^2}{I^2}} = \sqrt{\frac{Z}{Y}} = \sqrt{\frac{R+j\omega L}{G+j\omega C}}$$

4.5.1 The Lossy Case

The characteristic impedance for the case in which R and/or G are nonzero is referred to as the lossy case. The characteristic impedance raised to the second power is

$$Z_C^2 = \frac{Z}{Y} = \frac{R+j\omega L}{G+j\omega C}$$
(4-130)

Rationalizing the denominator and grouping terms into real and imaginary components yields the following:

$$Z_C^2 = \frac{RG - j\omega RC + j\omega LG - j^2\omega^2 LC}{G^2 + \omega C^2}$$

$$= \frac{RG + \omega^2 LC}{G^2 + (\omega C)^2} + j\omega\frac{LG - RC}{G^2 + (\omega C)^2}$$
(4-131)

This expression may be represented as a magnitude and an angle as follows:

$$|Z_C|^2 = \left\{ \left[\frac{RG+\omega^2 LC}{G^2+(\omega C)^2}\right]^2 + \left[\frac{\omega(LG-RC)}{G^2+(\omega C)^2}\right]^2 \right\}^{1/2}$$
(4-132)

$$\angle Z_C^2 = \tan^{-1}\left[\frac{\omega(LG-RC)}{RG+\omega^2 LC}\right]$$

The magnitude of Z_C is obtained by taking the square root of the magnitude of Z_C^2 as follows:

$$|Z_C| = \left\{ \left[\frac{RG+\omega^2 LC}{G^2+(\omega C)^2}\right]^2 + \left[\frac{\omega(LG-RC)}{G^2+(\omega C)^2}\right]^2 \right\}^{1/4}$$
(4-133)

CHAPTER 4

The angle of Z_C is obtained by taking the square root of the angle of Z_C^2 by dividing the angle by 2 as follows:

$$\angle Z_C = \frac{1}{2}\tan^{-1}\left[\frac{\omega(LG-RC)}{RG+\omega^2 LC}\right] \quad (4\text{-}134)$$

The characteristic impedance may now be summarized as follows:

$$|Z_C| = \left\{\left[\frac{RG+\omega^2 LC}{G^2+(\omega C)^2}\right]^2 + \left[\frac{\omega(LG-RC)}{G^2+(\omega C)^2}\right]^2\right\}^{1/4} \quad (4\text{-}135)$$

$$\angle Z_C = \frac{1}{2}\tan^{-1}\left[\frac{\omega(LG-RC)}{RG+\omega^2 LC}\right] \quad (4\text{-}136)$$

The real and imaginary components of the characteristic impedance are

$$R_0 = \Re Z_C = |Z_C|\cos\angle Z_C \quad (4\text{-}137)$$

$$X_0 = \Im Z_C = |Z_C|\sin\angle Z_C \quad (4\text{-}138)$$

Consider the following parameters for a 735-kV transmission line:

$$\begin{aligned} R &= 1.67\times 10^{-2} \text{ }\Omega/\text{mi} \\ L &= 1.40\times 10^{-3} \text{ H/mi} \\ C &= 2.05\times 10^{-8} \text{ F/mi} \\ G &= 1.00\times 10^{-11} \text{ Mhos/mi} \\ \omega &= 377 \text{ rad/s} \end{aligned} \quad (4\text{-}139)$$

The characteristic impedance is computed as follows:

$$|Z_C| = \left\{\left[\frac{RG+\omega^2 LC}{G^2+(\omega C)^2}\right]^2 + \left[\frac{\omega(LG-RC)}{G^2+(\omega C)^2}\right]^2\right\}^{1/4} = 261.13 \text{ }\Omega \quad (4\text{-}140)$$

$$\angle Z_C = \frac{1}{2}\tan^{-1}\left[\frac{\omega(LG-RC)}{RG+\omega^2 LC}\right] = -0.91° \quad (4\text{-}141)$$

TRANSMISSION LINES

The real and imaginary components are:

$$R_0 = \Re Z_C = |Z_C|\cos\angle Z_C = 261.10 \ \Omega \quad (4\text{-}142)$$

$$X_0 = \Im Z_C = |Z_C|\sin\angle Z_C = -4.15 \ \Omega \quad (4\text{-}143)$$

The characteristic impedance may also be computed from the series impedance and the shunt admittance as follows:

$$Z_C = \sqrt{\frac{R+j\omega L}{G+j\omega C}} = \sqrt{\frac{Z}{Y}} = \sqrt{\frac{|Z|}{|Y|}} \angle \frac{1}{2}(\theta_Z - \theta_Y) \quad (4\text{-}144)$$

4.5.2 The Lossless Case
Neglecting the series resistance and the shunt conductance yields the following expression for the characteristic impedance:

$$Z_C = \sqrt{\frac{L}{C}} \quad (4\text{-}145)$$

Substitution of the values cited previously yields

$$Z_C = 261.06 \angle 0° \quad (4\text{-}146)$$

It is interesting to note that the characteristic impedance is entirely real in the absence of losses. Note also that neglecting the series resistance and shunt conductance does not have a significant effect upon the magnitude of the characteristic impedance in this case.

4.6 PROPAGATION CONSTANT
The units of the propagation constant may be determined from the original expression as follows:

$$\gamma = \sqrt{(R+j\omega L)(G+j\omega C)} = \sqrt{(\Omega m^{-1})(\Omega m^{-1})} = \sqrt{m^{-2}} = m^{-1} \quad (4\text{-}147)$$

The propagation constant therefore has the dimension of m^{-1} or per-meter. This quantity is a complex number with a real and imaginary part. The propagation constant may be expressed in terms of the real and imaginary components as follows:

CHAPTER 4

$$\gamma = \alpha + \beta \qquad (4\text{-}148)$$

The real component α of the propagation constant represents the attenuation of the traveling wave per unit distance along the transmission line. This results from the presence of the series resistance and the leakage conductance.

The product of the real part α which is given the units of nepers[10] per meter and the distance along the transmission line x in meters yields a factor which, when applied to the magnitude of the traveling wave, determines the total attenuation in nepers at that point. The effect of the real component is to reduce the magnitude of the traveling wave as it travels along the length of the transmission line.

The imaginary component β represents the phase shift of the traveling wave per unit distance along the transmission line. The product of the imaginary part ß which is given in units of radians per meter and the distance along the transmission line x in meters yields a factor which, when applied to the phase of the traveling wave, determines the amount of phase shift. A single cycle of a sinusoidal wave which has traveled a distance of one wavelength, e.g., meters (m), will be shifted in phase by $\beta\lambda$ rad.

4.6.1 Lossy Case

The expression for the propagation constant for the case in which R, G, L, and C are all nonzero is restated for convenience as follows:

$$\gamma = \sqrt{ZY} = \sqrt{(R+j\omega L)(G+j\omega C)} \qquad (4\text{-}149)$$

$$\gamma^2 = ZY = (R+j\omega L)(G+j\omega C) \qquad (4\text{-}150)$$

Expanding and grouping terms into real and imaginary components yields the following:

$$\gamma^2 = RG + J\omega RC + j\omega LG + j^2\omega^2 LC \qquad (4\text{-}151)$$

---

[10]The unit neper is named in honor of the Scottish mathematician John Napier (1550-1617) who invented logarithms.

TRANSMISSION LINES

$$\gamma^2 = (RG - \omega^2 LC) + j\omega(LG + RC) \tag{4-152}$$

This expression may be represented in polar form with a magnitude

$$|\gamma^2| = \{(RG - \omega^2 LC)^2 + [\omega(LG + RC)]^2\}^{1/2} \tag{4-153}$$

The magnitude of γ is obtained by taking the square root of the

$$|\gamma| = \{(RG - \omega^2 LC)^2 + [\omega(LG + RC)]^2\}^{1/4} \tag{4-154}$$

The angle of γ is obtained by taking the square root of the angle of by dividing the angle by two as follows:

$$\angle \gamma = \frac{1}{2} \tan^{-1} \left[\frac{\omega(LG + RC)}{(RG - \omega^2 LC)} \right] \tag{4-155}$$

The propagation constant may now be summarized as follows:

$$|\gamma| = \{(RG - \omega^2 LC)^2 + [\omega(LG + RC)]^2\}^{1/4} \tag{4-156}$$

$$\angle \gamma = \frac{1}{2} \tan^{-1} \left[\frac{\omega(LG + RC)}{(RG - \omega^2 LC)} \right] \tag{4-157}$$

$$\alpha = \Re \gamma = |\gamma| \cos \gamma \tag{4-158}$$

$$\beta = \Im \gamma = |\gamma| \sin \gamma \tag{4-159}$$

Many of today's pocket calculators contain built-in functions for operations involving complex numbers. The values of R, L, G, and C may be entered into the appropriate registers, and the propagation constant may be obtained as the square root of the complex number multiplication of Z and Y. The result may then be expressed in rectangular form to obtain the real (α) and imaginary (β) components.

The transmission line parameters which were used in the calculation of the characteristic impedance for the 735-kV line are repeated as follows:

$$R=1.67\times10^{-2} \quad \Omega/\text{mi}$$
$$L=1.40\times10^{-3} \quad \text{H/mi}$$
$$C=2.05\times10^{-8} \quad \text{F/mi} \quad (4\text{-}160)$$
$$G=1.00\times10^{-11} \quad \text{Mhos/mi}$$
$$\omega=377 \quad \text{rad/s}$$

Substitution of these values into the expression for the propagation constant yields

$$\gamma = 8.34\times10^{-5}+j2.07\times10^{-3}$$
$$= 0.0020\angle 89.09° \quad (4\text{-}161)$$

The propagation constant may also be determined from the magnitudes and angles of the series impedance and the shunt admittance as follows:

$$\gamma = \sqrt{ZY} = \sqrt{|Z||Y|} \angle \frac{1}{2}(\theta_Z+\theta_Y) \quad (4\text{-}162)$$

4.6.2 The Lossless Case

The lossless case in which $R=G=0$ results in a propagation constant as follows:

$$\gamma = \sqrt{(0+j\omega L)(0+j\omega C)} = j\omega\sqrt{LC} \quad (4\text{-}163)$$

$$\alpha = 0 \quad (4\text{-}164)$$

$$\beta = \omega\sqrt{LC} \quad (4\text{-}165)$$

Note from earlier work that the velocity of propagation is

$$v = \sqrt{\frac{1}{LC}} \quad (4\text{-}166)$$

The phase coefficient β may then be expressed as follows:

$$\beta = \frac{\omega}{v} \quad (4\text{-}167)$$

This results in the following units:

$$\frac{\text{radians}}{\text{seconds}} \cdot \frac{\text{seconds}}{\text{mile}} = \frac{\text{radians}}{\text{mile}} \qquad (4\text{-}168)$$

The quantity βx there yields change in phase in radians over a given distance.

Note also that ω in the lossless case is the undamped natural frequency of the transmission line.

4.6.3 The Distortionless Case
The distortionless case is a special case of the lossy case in which the waveform maintains its shape as it propagates along the transmission line. This condition occurs when the following relationships exist among the line parameters:

$$\frac{G}{C} = \frac{R}{L} \qquad (4\text{-}169)$$

The requirements for the distortionless case are not consistent with the physical characteristics of power transmission lines.

4.7 A PRACTICAL EXAMPLE
The transmission line which is analyzed in this example is designed for a nominal system voltage of 735 kV. The line is 379 km in length or approximately 235 mi. Our intent is to determine the receiving-end voltage, current, and real and reactive power given the sending-end values.

The per-unit system was developed for power system calculations in order to simplify the analysis of networks in which more than one voltage exist. A set of base quantities are specified for each voltage. We will select a base voltage of 735 kV which is the nominal system voltage for this transmission line.

$$V_B = 735 \text{ kV} \qquad (4\text{-}170)$$

Consider that the voltage on the transmission line during the course of system operations is equal to the nominal system voltage of 735 kV.

CHAPTER 4

The per-unit (pu) voltage is then computed as the ratio of the actual system voltage to the base system voltage as follows:

$$V_{pu} = \frac{V}{V_B} = \frac{735}{735} = 1\,\text{pu} \qquad (4\text{-}171)$$

Now let us assume that the operating voltage has risen due to a reduction in load or some other factor to a new value of 750 kV. The new per-unit voltage is

$$V_{pu} = \frac{V}{V_B} = \frac{750}{735} = 1.02\,\text{pu} \qquad (4\text{-}172)$$

We have selected the voltage base as the nominal system voltage, and we will now select an apparent power base. It is a common practice in the industry to use 100 MVA as an apparent power base for power system calculations.

$$S_B = 100\ \text{MVA} \qquad (4\text{-}173)$$

Consider that the complex power flow from the sending terminal of the transmission line is

$$\begin{aligned} S &= P + jQ \\ &= 1400 - j400\ \text{MVA} \end{aligned} \qquad (4\text{-}174)$$

The per-unit power is computed as the ratio of the actual apparent power to the base apparent power as follows:

$$\begin{aligned} S_{pu} &= \frac{1400 - j400}{1000} \\ &= 14 - j4\ \text{pu} \end{aligned} \qquad (4\text{-}175)$$

Now let us consider the line parameters. We must first compute the base impedance from the base voltage and the base apparent power as follows:

$$Z_B = \frac{V_B^2}{S_B} = \frac{735^2}{100} = 5402.25 \qquad (4\text{-}176)$$

In the case of the voltage and apparent power we computed the per-unit quantities from the actual quantities and the base quantities. We will now compute the actual quantities for a set of transmission line data from the per-unit quantities and the base quantities.

Table 4-3 Line Data

R	X	B
0.00073	0.02294	9.8232

The per-unit series resistance is R, the per-unit series reactance is X, and the per-unit shunt susceptance is B. Note that the shunt conductance is not included in the line data given above since it does not significantly affect the results of load flow calculations.[11]

It is worthwhile to point out a note of interest about the shunt susceptance parameter. The line charging at nominal voltage is the amount of reactive power associated with the application of the nominal system voltage to the shunt capacitance of the line. Consider that the shunt susceptance is given as follows:

$$B = j\omega C \qquad (4\text{-}177)$$

The line charging may be computed as follows:

$$Q = V^2 B \qquad (4\text{-}178)$$

The computation of the line charging at nominal voltage on a per-unit basis yields

$$9.8232 = (1)^2 (9.8232) \qquad (4\text{-}179)$$

Note that the line charging at nominal voltage is numerically equal to the shunt susceptance based on per-unit quantities. We will now compute the actual line data from the per-unit quantities and the base

---

[11] The reader must not assume that the shunt conductance is not an important quantity. Power factor testing is conducted on a regular basis to monitor the level of the shunt conductance through the insulation of power system devices. A significant increase in the power factor may be indicative of a degradation of the insulation and may require the removal of the device from service for further testing.

quantities. The actual impedance is computed from the per-unit impedance and the base impedance as follows:

$$
\begin{aligned}
Z &= Z_{pu} \cdot Z_B \\
&= (R_{pu} + jX_{pu})Z_B \\
&= (0.00073 + j0.02294)(5402.25) \\
&= 3.94 + j123.93 \\
&= 123.99 \angle 88.18° \ \Omega
\end{aligned}
\qquad (4\text{-}180)
$$

The actual impedance per mile is

$$
\frac{Z}{l} = \frac{123.99 \angle 88.18°}{235} = 0.53 \angle 88.18° \ \Omega/\text{mi} \qquad (4\text{-}181)
$$

The actual series resistance is therefore 3.94 Ω, and the series reactance is 123.93 Ω. The series inductance can be computed from the series inductive reactance and the radian frequency as follows:

$$
L = \frac{X}{\omega} = \frac{\omega L}{\omega} = \frac{123.93}{377} = 0.33 \ \text{H} \qquad (4\text{-}182)
$$

The series resistance per mile is

$$
\frac{R}{l} = \frac{3.94}{235} = 1.67 \times 10^{-2} \ \Omega/\text{mi} \qquad (4\text{-}183)
$$

The series reactance per mile is

$$
\frac{X}{l} = \frac{123.93}{235} = 0.53 \ \Omega/\text{mi} \qquad (4\text{-}184)
$$

The series inductance per mile is

$$
\frac{L}{l} = 1.40 \times 10^{-3} \ \text{H/mi} \qquad (4\text{-}185)
$$

The actual value of the shunt admittance is computed from the per-unit shunt admittance and the base shunt admittance. The base admittance is the reciprocal of the base impedance.

$$
Y_B = \frac{1}{Z_B} = \frac{1}{5402.25} = 0.00019 \ \text{mhos} \qquad (4\text{-}186)
$$

TRANSMISSION LINES

The shunt conductance is neglected for this analysis, and the shunt admittance consists only of the shunt susceptance.

$$Y = Y_{pu} \cdot Y_B$$
$$= (j9.8232)(0.00019) = j0.001818 \text{ mhos} \qquad (4\text{-}187)$$

The shunt capacitance is

$$C = \frac{B}{\omega} = \frac{\omega C}{\omega} = \frac{0.001818}{377} = 4.82 \times 10^{-6} \text{ F} \qquad (4\text{-}188)$$

The shunt susceptance per mile is

$$\frac{B}{l} = \frac{0.001818}{235} = 7.72 \times 10^{-6} \text{ mhos/mi} \qquad (4\text{-}189)$$

The shunt capacitance per mile is

$$\frac{C}{l} = \frac{4.82 \times 10^{-6}}{235} = 2.05 \times 10^{-8} \text{ F/mi} \qquad (4\text{-}190)$$

The reader will recognize the series resistance per mile, the series inductance per mile, and the shunt capacitance per mile as the values that were previously used to compute the characteristic impedance and the propagation constant γ. The characteristic impedance Z_C is

$$Z_C = \sqrt{\frac{Z}{Y}} = 261.13 \angle -0.91° \qquad (4\text{-}191)$$

The propagation constant γ is

$$\gamma = \sqrt{ZY} = 0.0020 \angle 89.09° \qquad (4\text{-}192)$$

The term γl is

$$\gamma l = 235(0.0020 \angle 89.09°)$$
$$= 0.47 \angle 89.09° \qquad (4\text{-}193)$$
$$= 0.0076 + j0.4748$$

We will now use the foregoing information to solve for the receiving-end voltage, current, and real and reactive power given the corresponding sending-end quantities. We will assume that the phase-to-phase voltage at the sending-end terminal is 1.00 or 735 kV.

100   CHAPTER 4

The corresponding phase-to-ground voltage is then

$$\frac{735}{\sqrt{3}} = 424 \text{ kV} \qquad (4\text{-}194)$$

We will assume that the sending end voltage is our reference so that the voltage angle of the sending end voltage is zero.

$$V_S = 424 \angle 0° \qquad (4\text{-}195)$$

The complex power at the sending-end terminal is the complex sum of the real power flow in megawatts and the reactive power flow in megavars. The following three-phase values will be assumed:

$$P_S = 1400 \text{ MW} \qquad (4\text{-}196)$$

$$Q = -j400 \text{ MVAR} \qquad (4\text{-}197)$$

The sign conventions for the real and reactive power are such that the real power of +1400 MW is flowing from the sending bus into the transmission line, and the reactive power of -400 MVAR is flowing in the opposite direction, e.g., from the transmission line into the sending bus. The complex power is

$$\begin{aligned} S_S &= P_S + jQ_S \\ &= 1400 - j400 \\ &= 1456 \angle -15.9° \text{ MVA} \end{aligned} \qquad (4\text{-}198)$$

The complex power may also be expressed as follows:

$$S_S = \sqrt{3} V_S \hat{I}_S \qquad (4\text{-}199)$$

The conjugate of the phase current at the sending-end terminal is

$$\hat{I}_S = \frac{S_S}{\sqrt{3} V_S} = \frac{1456 \angle -15.9°}{\sqrt{3}(735 \angle 0°)} = 1.14 \angle -15.9° \text{ kA} \qquad (4\text{-}200)$$

Reversing the sign to obtain the actual phase current yields

$$I_S = 1.14 \angle 15.9° \qquad (4\text{-}201)$$

The voltage and current solutions were derived previously and are repeated here.

TRANSMISSION LINES

$$V_R = AV_S - BI_S \tag{4-202}$$

$$I_R = -CV_S + DI_S \tag{4-203}$$

The transmission line constants $ABCD$ are

$$A = \cosh(\gamma l) \qquad B = Z_C \sinh(\gamma l)$$

$$C = \frac{1}{Z_C}\sinh(\gamma l) \qquad D = \cosh(\gamma l)$$

We must introduce the following identities to solve the hyperbolic functions for voltage and current with the complex argument $\gamma l = \alpha l + j\beta l$.

$$\sinh(u+jv) = \sinh u \cos v + j\cosh u \sin v \tag{4-204}$$

$$\cosh(u+jv) = \cosh u \cos v + j\sinh u \sin v \tag{4-205}$$

The hyperbolic terms of the voltage and current equations may therefore be expressed as follows:

$$\sinh(\gamma l) = \sinh(\alpha l + j\beta l) = \sinh(\alpha l)\cos(\beta l) + j\cosh(\alpha l)\sin(\beta l) \tag{4-206}$$

$$\cosh(\gamma l) = \cosh(\alpha l + j\beta l) = \cosh(\alpha l)\cos(\beta l) + j\sinh(\alpha l)\sin(\beta l) \tag{4-207}$$

Substituting the values for α, β, and l to compute $\sinh(\gamma l)$ yields

$$\begin{aligned}\sinh(\gamma l) &= \sinh(0.0076 + j0.4748) \\ &= \sinh(0.0076)\cos(0.4748) + j\cosh(0.0076)\sin(0.4748) \\ &= 0.0067 + j0.4748 \\ &= 0.46\angle 89.16°\end{aligned} \tag{4-208}$$

Substituting the values for α, β, and l to compute $\cosh(\gamma l)$ yields

$$\begin{aligned}\cosh(\gamma l) &= \cosh(0.0076 + j0.4748) \\ &= \cosh(0.0076)\cos(0.4748) + j\sinh(0.0076)\sin(0.4748) \\ &= 0.8894 + j0.0035 \\ &= 0.89\angle 0.22°\end{aligned} \tag{4-209}$$

CHAPTER 4

The transmission line constants $ABCD$ are then

$$A = 0.89 \angle 0.22° \qquad B = 119.39 \angle 88.25°$$
$$C = 0.0018 \angle 90.07° \qquad D = 0.89 \angle 0.22°$$

The receiving-end voltage is computed as follows:

$$V_R = AV_S - BI_S$$
$$= (0.89 \angle 0.22°)(424 \angle 0°) - (119.39 \angle 88.25°)(1.14 \angle 15.9°) \qquad (4\text{-}210)$$
$$= 431.3 \angle -17.7° \text{ kV}$$

This translates to a phase-to-phase voltage at the receiving end of

$$(431.3 \angle -17.7°)(\sqrt{3}) = 747.0 \angle -17.7° \qquad (4\text{-}211)$$

The receiving-end current is computed as follows:

$$I_R = -CV_S = DI_S$$
$$= -(0.0018 \angle 89.99°)(424 \angle 0°) + (0.89 \angle 0.22°)(1.14 \angle 15.9° \qquad (4\text{-}212)$$
$$= 1.08 \angle -25.12° \text{ kA}$$

The receiving-end apparent power is computed as follows:

$$S_R = \sqrt{3} V \hat{I}$$
$$= \sqrt{3}(747.0 \angle -17.7°)(1.08 \angle 25.18) \qquad (4\text{-}213)$$
$$= 1398 \angle 7.5° \text{ MVA}$$
$$= 1386 \text{ MW} + j183 \text{ MVAR}$$

The reader is encouraged to perform the same analysis with the equivalent and approximate models developed previously to compare the results. The following points are significant:

1. The sending-end real power is 1400 MW and the receiving-end power is 1386 MW which yields a 14 MW loss in the resistance of the line.

2. The line contributes 400 MVAR of reactive power to the sending-end bus and 183 MVAR of reactive power to the receiving bus.

3. The magnitude of the receiving-end voltage of 747 kV exceeds the magnitude of the sending-end voltage of 735 kV which is consistent with the direction of reactive power flow. The

TRANSMISSION LINES   103

sending-end voltage angle leads the receiving-end voltage by 17.6° which is consistent with the direction of real power flow. We will examine these effects in Chap. 6 in our discussion of power circle diagrams.

4. The current at the sending-end bus of 1.14 kA and the current at the receiving-end bus of 1.08 kA lead the respective bus voltages since reactive power is flowing into both buses.

We have completed our analysis of a transmission line which is representative of the longest lines in North America. We will examine transmission lines in Chaps. 5 and 6 which are representative of bulk power lines under 100 mi in length.

CHAPTER 5

POWER SYSTEM LOADS

5.1 LOAD BUS SPECIFICATIONS

Let us consider the addition of a load bus to the power system. Service classifications which are assigned by the electric utilities include residential, commercial, light industrial, and heavy industrial loads as well as municipal electric company loads. We will first prepare a load bus specification which is a summary of the service requirements that must be provided by the electric utility. Let us assume that the total new load which is connected to the system is 200 MVA at 0.8 power factor lagging. Let us further assume that the service will be provided at the transmission level from a radial 115-kV circuit. Remember that the reactive power is positive for lagging power factor in accordance with standard conventions.

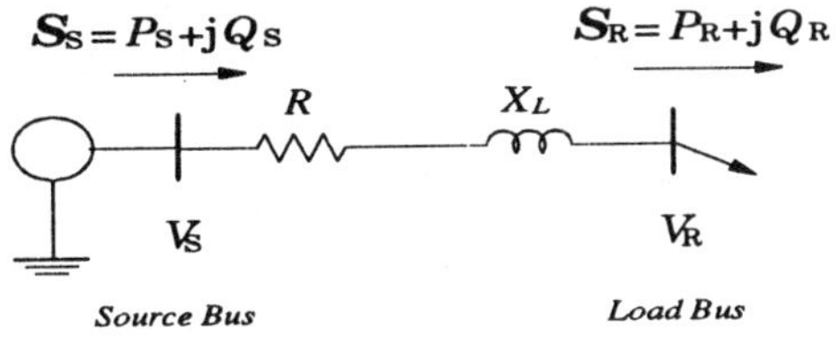

Figure 5-1
Operating system.

The three-phase apparent power can be represented in rectangular form as the vector sum of the real and reactive power components.

$$S = P + jQ \tag{5-1}$$

where S is the apparent power in megavolt-amperes, P is the real power in megawatts, and Q is the reactive power in megavars.

CHAPTER 5

The real and reactive power components can be computed as

$$P = \sqrt{S^2 - Q^2} \qquad (5\text{-}2)$$

$$Q = \sqrt{S^2 - P^2} \qquad (5\text{-}3)$$

The apparent power may also be represented in polar form with a magnitude and an angle.

$$|S| = \sqrt{P^2 + Q^2} \qquad (5\text{-}4)$$

$$\theta = \tan^{-1}\left(\frac{Q}{P}\right) \qquad (5\text{-}5)$$

The real and reactive power components can then be represented as

$$P = S\cos\theta \qquad (5\text{-}6)$$

$$Q = S\sin\theta \qquad (5\text{-}7)$$

The term $\cos\theta$ is the power factor which is the ratio of the real power component to the apparent power. The real power for this example can then be computed from the apparent power and the power factor which are given in the load bus specification.

$$\begin{aligned} P &= S\cos\theta \\ &= S\left(\frac{P}{S}\right) \\ &= (200)(0.8) = 160 \text{ MW} \end{aligned} \qquad (5\text{-}8)$$

The phase angle is computed from the arccosine of the power factor as follows:

$$\theta = \cos^{-1} 0.8 = 36.9° \qquad (5\text{-}9)$$

The reactive power is computed as follows:

$$\begin{aligned} Q &= \sqrt{S^2 - P^2} \\ &= \sqrt{200^2 - 160^2} \\ &= 120 \text{ MVAR} \end{aligned} \qquad (5\text{-}10)$$

POWER SYSTEM LOADS

The load impedance is computed as follows:

$$Z = \frac{V^2}{S} = \frac{(115)^2}{200} = 66.1 \ \Omega \quad (5\text{-}11)$$

The load resistance is computed as follows:

$$R = Z\cos\theta = 66.1\cos 36.9° = 52.9 \ \Omega \quad (5\text{-}12)$$

The load reactance is computed as follows:

$$X_L = Z\sin\theta = 66.1\sin 36.9° = 39.7 \ \Omega \quad (5\text{-}13)$$

The load current can be obtained from an alternate expression for the apparent power.

$$S = \sqrt{3} \, V\hat{I} \quad (5\text{-}14)$$

Table 5-1 Load Bus Data

QUANTITY	EXPRESSION	VALUE
Voltage	V	115 kV
Current	I	1 kA
Apparent power	S	200 MVA
Real power	P	160 MW
Reactive power	Q	120 MVAR
Power factor	$\cos\theta$	0.8 lagging
Phase angle	θ	36.9°
Load impedance	Z	66.1 Ω
Load resistance	R	52.9 Ω
Load reactance	X_L	39.7 Ω

108   CHAPTER 5

The conjugate of the load current is

$$\hat{I} = \frac{S}{\sqrt{3}\,V} = \frac{200\angle 36.9°}{\sqrt{3}(115\angle 0°)} = 1 \text{ kA}\angle 36.9° \qquad (5\text{-}15)$$

The load current is

$$I = 1 \text{ kA}\angle -36.9° \qquad (5\text{-}16)$$

Note that a positive power angle represents a lagging condition so that the load current lags the applied voltage by 36.9°. The load bus specifications are summarized in Table 5-1.

5.2 TRANSMISSION LINE SPECIFICATIONS

Consider that the load will be served from a 115-kV transmission line. The line has the following resistance and inductive reactance.

$$R = 3 \; \Omega \qquad (5\text{-}17)$$

$$X_L = 8 \; \Omega \qquad (5\text{-}18)$$

The line impedance is computed as follows:

$$\begin{aligned} Z &= \sqrt{R^2 + X_L^2} \\ &= \sqrt{3^2 + 8^2} \\ &= 8.5 \; \Omega \end{aligned} \qquad (5\text{-}19)$$

The line impedance angle is computed as follows:

$$\theta = \tan^{-1}\left(\frac{X_L}{R}\right) = \tan^{-1}\left(\frac{8}{3}\right) = 69.4° \qquad (5\text{-}20)$$

The real power loss for each of the three phases which is the amount of real power that is dissipated by the flow of current through the resistance of the phase conductors is

$$P_{\text{loss}} = I^2 R = (1)^2(3) = 3 \text{ MW} \qquad (5\text{-}21)$$

The total real power loss in the transmission line is then 9 MW. The reactive power loss per phase which is the amount of reactive power that is absorbed by the flow of current through the inductive reactance

POWER SYSTEM LOADS

Table 5-2 TRANSMISSION LINE DATA

QUANTITY	EXPRESSION	VALUE
Resistance	R	3 Ω
Inductive reactance	X_L	8 Ω
Series impedance	Z	8.5 Ω
Series impedance angle	θ	69.4°
Real power loss	P_{loss}	9 MW
Reactive power loss	Q_{loss}	24 MVAR
Voltage drop	V_{drop}	8.5 kV

of the phase conductors is

$$Q_{loss}=I^2X_L=(1)^2(8)=8 \text{ MVAR} \qquad (5\text{-}22)$$

The total reactive power loss in the transmission line is then 24 MVAR. The voltage drop between the source bus and the load bus which results from the flow of load current through the series impedance of the transmission line is computed as follows:

$$V_{drop}=IZ=(1)(8.5)=8.5 \text{ kV} \qquad (5\text{-}23)$$

The transmission line data are summarized in Table 5-2.

5.3 SOURCE BUS SPECIFICATIONS

The voltage at the source bus will equal the sum of the voltage at the load bus and the voltage drop along the transmission line.

$$V_S=IZ+V_R \qquad (5\text{-}24)$$
$$=8.5+115=123.5 \text{ kV}$$

The real power delivered from the source bus is the sum of the real power loss in the line and the real power load at the load bus.

$$P=P_{loss}+P_R \qquad (5\text{-}25)$$
$$=9+160=169 \text{ MW}$$

CHAPTER 5

Table 5-3 SOUCE BUS DATA

QUANTITY	LOAD BUS	LINE	SOURCE BUS
Voltage	115	8.5	123.5 kV
Real power	160	9	169 MW
Reactive power	120	24	144 MVAR
Apparent power	200	—	222 MVA
Angle	36.9°	—	40.4°

The reactive power delivered from the source bus is the sum of the reactive power loss in the line and the reactive power load at the load bus.

$$Q_S = Q_{loss} + Q_R \qquad (5\text{-}26)$$
$$= 24 + 120 = 144 \text{ MVAR}$$

The apparent power at the source bus is then the vector sum of the real and reactive power components.

$$S = \sqrt{P^2 + Q^2} = 207.3 \text{ MVA} \qquad (5\text{-}27)$$

The apparent power angle at the source bus is

$$\theta = \tan^{-1}\left(\frac{Q}{P}\right) = 38.1° \qquad (5\text{-}28)$$

The source bus specifications are summarized in Table 5-3.

5.4 POWER FACTOR CORRECTION

The operating efficiency and voltage regulation of the transmission line can be improved by providing reactive power support at the load. This precludes the requirement to transmit the reactive power requirements for the load through the transmission line. The amount of reactive power that is needed to provide 100 percent compensation is 120 MVAR. We will add a three-phase shunt capacitor bank with a nominal voltage rating of 115 kV and a reactive power rating at a nominal voltage of 120 MVAR. The capacitor bank may be switchable

POWER SYSTEM LOADS

as illustrated in Fig. 5-2 to remove the source of reactive power during light-load periods or hard-wired to the load bus. The reduction in the megavoltampere level at the load results in a reduction in the load current and hence the current through the transmission line. This in turn results in a reduction in the real and reactive power losses in the line and a reduction in the voltage drop between the sending and receiving buses. The impedance of the bank is computed as follows:

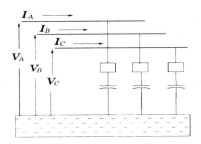

Figure 5-2 Capacitor bank.

$$Z = \frac{V^2}{Q} = \frac{115^2}{120} = 110.2 \ \Omega \quad (5\text{-}29)$$

Table 5-4 Load Bus Data with Power Factor Correction

QUANTITY	EXPRESSION	VALUE
Voltage	V	115 kV
Current	I	0.8 kA
Apparent power	S	160 MVA
Real power	P	160 MW
Reactive power	Q	0 MVAR
Power factor	$\cos\theta$	1
Phase angle	θ	0°
Load impedance	Z	82.7 Ω
Load resistance	R	82.7 Ω
Load reactance	X_L	0 Ω

112   CHAPTER 5

The load bus specifications are modified as in Table 5-4. Note that the current flow is now reduced from 1 to 0.8 kA due to the reduction in the apparent power requirement at the load from 200 to 160 MVA. The entire reactive power requirement of 120 MVA has been satisfied by the application of the shunt capacitor bank.

The real power loss for each of the three phases is reduced from 3 MW per phase to

$$P_{loss} = I^2 R = (0.8)^2(3) = 1.9 \text{ MW} \tag{5-30}$$

The total real power loss in the transmission line for the three phases is then 5.7 MW. The reactive power loss per phase is reduced from 8 MVAR to

$$Q_{loss} = I^2 X_L = (0.8)(8) = 5.1 \text{ MVAR} \tag{5-31}$$

The total reactive power loss in the transmission line is then reduced from 24 to 15.3 MVAR. The voltage drop between the source bus and the load bus which results from the flow of load current through the series impedance of the transmission line is reduced from 8.5 kV to

$$V_{drop} = IZ = (0.8)(8.5) = 6.8 \text{ kV} \tag{5-32}$$

5.5 TRANSMISSION SYSTEM EFFICIENCY

The operating efficiency of a transmission line is the ratio of the real power delivered to the receiving-end bus to the real power transferred from the sending-end bus. The difference between these two quantities is the real power loss in the transmission line. Consider the previous example. The transmission line efficiency without power factor correction is computed as follows:

$$\eta = \frac{P_{load}}{P_{source}} \times 100\%$$
$$= \frac{160}{169} \times 100 \tag{5-33}$$
$$= 94.7 \%$$

The transmission line efficiency with power factor correction is

POWER SYSTEM LOADS

$$\eta = \frac{P_{load}}{P_{source}} \times 100\%$$
$$= \frac{160}{165.8} \times 100 \quad (5\text{-}34)$$
$$= 96.5\ \%$$

An upgrade of the transmission system which results in a doubling of the operating voltage permits a reduction of phase currents to one-half the previous value for the same level of apparent power transfer. The three-phase expression for apparent power as a function of the phase-to-phase voltage and the phase current is restated as follows:

$$S = \sqrt{3}\,VI \quad (5\text{-}35)$$

An increase in the operating voltage to twice the initial value and a corresponding reduction of the phase currrent to one-half the initial value yields the same level of apparent power.

$$S = \sqrt{3}\,(2V)\left(\frac{1}{2}I\right) = \sqrt{3}\,VI \quad (5\text{-}36)$$

The reduction of the phase current to one-half the previous value results in a reduction of the real and reactive power losses to one-fourth the previous value.

$$P_{loss} = \left(\frac{1}{2}I\right)^{2} R = \frac{1}{4}I^{2}R \quad (5\text{-}37)$$

$$Q_{loss} = \left(\frac{1}{2}I\right)^{2} X_{L} = \frac{1}{4}I^{2}X_{L} \quad (5\text{-}38)$$

The voltage drop across the impedance of the transmission system is also reduced to one-half the previous value as follows:

$$V_{drop} = \left(\frac{1}{2}I\right) Z = \frac{1}{2}IZ \quad (5\text{-}39)$$

5.6 TRANSMISSION LINE VOLTAGE REGULATION

The transmission line voltage regulation (TLVR) is the ratio of the voltage drop between the sending-end and receiving-end buses to the receiving-end voltage. The nominal system voltage is sometimes used instead of the sending-end voltage. The TLVR for the previous example without power factor correction is

$$\text{TLVR} = \frac{V_{drop}}{V_{nominal}} \times 100\%$$
$$= 8.\frac{5}{115} \times 100\% \quad (5\text{-}40)$$
$$= 7.4\%$$

The TLVR in this case indicates that the voltage drop between the sending and receiving buses is maintained to within 7.4% of the nominal system voltage.

The TLVR with power factor correction is improved to

$$\text{TLVR} = \frac{8.5}{115} \times 100\% \quad (5\text{-}41)$$
$$= 5.9\%$$

5.7 LOAD FACTOR

The two components of electric service are demand and energy. Demand is the maximum level of real power which the electric utility must supply to satisfy the load requirements of its customers. Energy is the cumulative use of electric power over a period of time. The demand component of the electric rate structure represents the capital investment required of the electric utility to provide the generation, transmission, and distribution facilities needed to meet the maximum concurrent level of customer demand.

The load factor is the ratio of the actual energy usage to the rated maximum energy usage over a given period. Utilities often define load periods in terms of on-peak and off-peak hours. On-peak hours may extend from 6:00 am to 10:00 pm which is the period of highest demand or the peak load period. The off-peak hours extend over the balance of the day which is generally the period of lowest demand or the light-load period. The actual peak-load and light-load periods are, of course, dependent upon the nature of the load.

POWER SYSTEM LOADS 115

A manufacturing facility, for example, which operates three shifts over a 24-h period may have nearly the same demand during the on-peak and off-peak periods. An urban subway system, on the other hand, which provides the major portion of commuter service during the morning and evening rush hours, may have a peak demand during those periods and lower demand during the balance of the day.

An economic incentive is offered which involves lower electric rates for the sale of off-peak energy in order to levelize the demand by diverting a portion of the energy usage from the on-peak to the off-peak periods. Time-of-use metering may be employed which assigns a given cost per kilowatt-hour to on-peak loads and a lower cost per kilowatt-hour for off-peak loads.

A low load factor is indicative of a substantial period during which the capacity of the system is underutilized. More efficient utilization of the installed capacity is realized when the system loads do not occur concurrently, e.g., when the disparity between on-peak and off-peak demand is reduced. An 11-month demand profile in kilowatts for an industrial organization is illustrated in the bar chart of Fig. 5-3. Note that the three shifts of manufacturing result in a fairly level demand between the on-peak and off-peak hours.

The energy component represents the operating costs which include fuel, maintenance, and similar items that must be provided to meet the demand requirements over a given period.

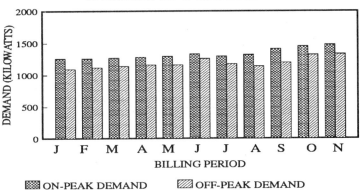

Figure 5-3 Demand.

ENERGY PROFILE

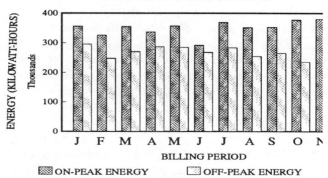

Figure 5-4 Energy.

The energy usage for the same organization is illustrated in Fig. 5-4. We have seen in this chapter that the delivery of reactive power requirements over the transmission system results in real and reactive power losses which reduce the operating efficiency of the system. Electric utilities may therefore assess penalties for the delivery of reactive power when the power factor of the customer's load is below a minimum level such as 90 percent. The reactive energy usage of the manufacturing organization is illustrated in Fig. 5-5.

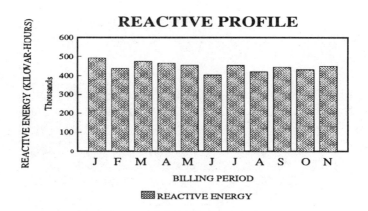

Figure 5-5 Reactive energy.

CHAPTER 6

POWER CIRCLE DIAGRAMS

Power circle diagrams convey a great deal of information about the operation of a transmission line. This device is a graphical representation of the real and reactive power flow at the sending and receiving terminals versus the magnitudes and angles of the terminal voltages. The advantage of understanding power circle diagrams lies in the ability to visualize the performance of a transmission line in response to changing conditions at the line terminals. This tool is particularly useful to the system operations engineer and the power dispatcher who make decisions regarding the switching of reactive devices, changing of transformer taps, and loading of the transmission system.

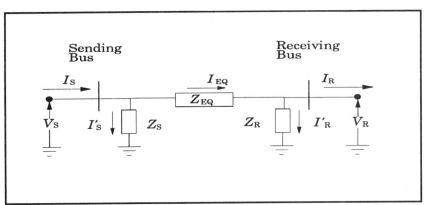

Figure 6-1 Transmission model.

One terminal of the transmission line is designated as the sending bus, and the other terminal is designated as the receiving bus. In applying power circle diagrams to the analysis of operating transmission lines, the sending bus may represent a source bus and the receiving bus may represent a load bus. These assignments may be made arbitrarily, however, in cases where the power flow may occur in either direction

118 CHAPTER 6

on a given transmission line. The power circle diagram is based on the approximate pi model. The series impedance Z_{EQ} is computed from the phasor sum of the series resistance R and the series inductive reactance X_L, and the shunt impedances at the sending and receiving buses Z_S and Z_R are based on the shunt capacitive reactance. The effect of the shunt conductance G is insignificant in terms of transmission line operation and is therefore neglected.

6.1 DERIVATION OF THE POWER CIRCLE DIAGRAM EQUATIONS

The power circle diagram equations may be derived on the basis of the current flow into the sending and receiving buses. The current flow from the sending bus into the line is equal to the sum of the current flow into the series impedance Z_{EQ} and the current flow into the shunt impedance Z_S; similarly, the current flow into the receiving bus from the line is equal to the current flow from the series equivalent impedance Z_{EQ} and the current flow into the shunt impedance Z_R at the receiving bus. The relationship for the sending bus may be stated as follows:

$$\hat{I}_S = \hat{I}_{EQ} + \hat{I}_S' \qquad (6\text{-}1)$$

Note that the conjugate of the sending current is used to be consistent with the standard convention. The above equation may be restated to represent the current as the ratio of the sending and receiving bus voltages to the series and shunt impedances.

$$\hat{I}_S' = \frac{\hat{V}_S}{\hat{Z}_S} \qquad (6\text{-}2)$$

$$\hat{I}_{EQ} = \frac{\hat{V}_S - \hat{V}_R}{\hat{Z}_{EQ}} \qquad (6\text{-}3)$$

$$\hat{I}_S = \frac{\hat{V}_S}{\hat{Z}_S} + \frac{\hat{V}_S - \hat{V}_R}{\hat{Z}_{EQ}} \qquad (6\text{-}4)$$

POWER CIRCLE DIAGRAMS 119

Rearranging terms,

$$\hat{I}_S = \frac{\hat{V}_S}{\hat{Z}_S} + \frac{\hat{V}_S}{\hat{Z}_{EQ}} - \frac{\hat{V}_R}{\hat{Z}_{EQ}} \quad (6\text{-}5)$$

The power circle diagram for the sending bus may now be developed as the product of the sending bus current given in (6-5) and the sending bus voltage.

$$V_S \hat{I}_S = \frac{V_S \hat{V}_S}{\hat{Z}_S} + \frac{V_S \hat{V}_S}{\hat{Z}_{EQ}} - \frac{V_S \hat{V}_R}{\hat{Z}_{EQ}} \quad (6\text{-}6)$$

It should be noted that any phasor multiplied by its conjugate is equal to the square of the magnitude of the phasor. The first and second terms therefore become

$$\frac{V_S \hat{V}_S}{\hat{Z}_S} + \frac{V_S \hat{V}_S}{\hat{Z}_{EQ}} = \frac{V_S^2}{\hat{Z}_S} + \frac{V_S^2}{\hat{Z}_{EQ}} \quad (6\text{-}7)$$

In addition, the receiving voltage is taken as the reference voltage so that the voltage angle at the receiving bus is equal to zero.

$$V_R = \hat{V}_R = V_R \angle 0° = V_R \quad (6\text{-}8)$$

The third term becomes

$$\frac{V_S \hat{V}_R}{\hat{Z}_{EQ}} = \frac{V_R V_S \varepsilon^{+j\theta_v}}{\hat{Z}_{EQ}} \quad (6\text{-}9)$$

The sending voltage is now expressed in exponential form. The voltage angle θ_V is the difference between the sending voltage angle and the receiving voltage angle. Note that the voltage angle at the receiving bus is taken as the reference voltage angle and is therefore assumed to be zero. It should also be noted that the voltage angle θ_V between the sending and receiving bus voltages is not to be confused with the phase angle θ between voltage and current. The final equation for sending power is

$$P_S + jQ_S = \frac{V_S^2}{\hat{Z}_S} + \frac{V_S^2}{\hat{Z}_{EQ}} - \frac{V_R V_S \varepsilon^{j\theta_v}}{\hat{Z}_{EQ}} \quad (6\text{-}10)$$

The power circle diagram equation for the receiving bus may be developed in the same manner as for the sending bus which results in the following equation:

$$P_R + jQ_R = -\frac{V_R^2}{\hat{Z}_R} - \frac{V_R^2}{\hat{Z}_{EQ}} + \frac{V_R V_S \varepsilon^{-j\theta_v}}{\hat{Z}_{EQ}} \qquad (6\text{-}11)$$

6.2 A PRACTICAL EXAMPLE OF REAL AND REACTIVE POWER FLOW

The following example is based on the operation of a typical 345-kV transmission line. The electrical parameters for this line are listed in Table 6-1. The power circle diagram equation for the sending bus is restated as follows:

$$P_S + jQ_S = \frac{V_S^2}{\hat{Z}_S} + \frac{V_S^2}{\hat{Z}_{EQ}} - \frac{V_R V_S \varepsilon^{j\theta_v}}{\hat{Z}_{EQ}} \qquad (6\text{-}12)$$

Table 6-1 Transmission Line Parameters

QUANTITY	EXPRESSION	VALUE
Sending voltage	V_S	350 kV
Receiving voltage	V_R	350 kV
Resistance	R	3 Ω
Inductive reactance	X_L	42 Ω
Shunt capacitive reactance	X_C	1920 Ω
Series impedance	Z_{EQ}	42.1 Ω
Series impedance angle	θ	85.9°
Shunt impedance	Z_S, Z_R	3840 Ω

POWER CIRCLE DIAGRAMS 121

The three phasors on the right side of the equation are computed as follows:

$$\frac{V_S^2}{Z_S} = 31.9 \angle 90° \text{ MVA} \qquad (6\text{-}13)$$

$$\frac{V_S^2}{Z_{EQ}} = 2910 \angle 85.9° \text{ MVA} \qquad (6\text{-}14)$$

$$-\frac{V_S V_R \varepsilon^{j\theta_V}}{Z_{EQ}} = 2910 \angle (-85.9° + \theta_V) \text{ MVA} \qquad (6\text{-}15)$$

The graphical addition of the three phasors for the sending bus is illustrated in Fig. 6-2. The first and second phasors are fixed by the sending voltage V_S, the sending bus shunt impedance Z_S, and the series equivalent impedance Z_{EQ}. The third phasor has an initial angular position which is based on the angle of the series equivalent impedance. The voltage angle θ_V is then added to determine the associated values of real and reactive power at the sending bus.

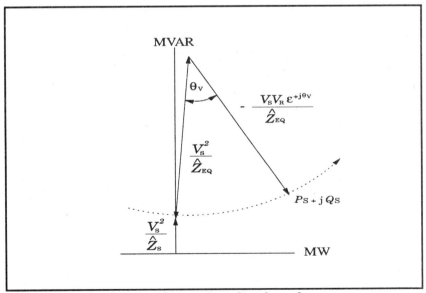

Figure 6-2 Graphical addition of sending bus phasors.

122   CHAPTER 6

A full plot of the phasors through 360° yields the power circle diagram for the sending bus.

The graphical addition of the three phasors for the receiving bus is illustrated in Fig. 6-3. The first and second phasors are fixed by the receiving voltage V_R, the receiving bus shunt impedance Z_R and the series equivalent impedance Z_{EQ}. The third phasor has an initial angular position which is based on the angle of the series equivalent impedance. The voltage angle θ_V is then added to determine the associated values of real and reactive power at the receiving bus. A full plot of the phasors through 360° yields the power circle diagram for the receiving bus.

A plot of the power circle diagrams for the 345-kV transmission line is illustrated in Fig. 6-4. This plot was developed from the actual line data with the sending and receiving bus voltages held constant at 350 kV. The voltage angle was then incremented in graduations of 10° to provide 36 points of apparent power for the sending and receiving buses. The reader can determine from this diagram that the maximum power transfer occurs when the voltage angle approaches 90° and that further increases in voltage angle will result in a reduction in the real power transmitted from the sending bus and

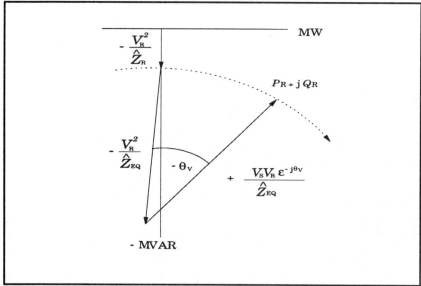

Figure 6-3 Graphical addition of receiving bus phasors.

POWER CIRCLE DIAGRAMS 123

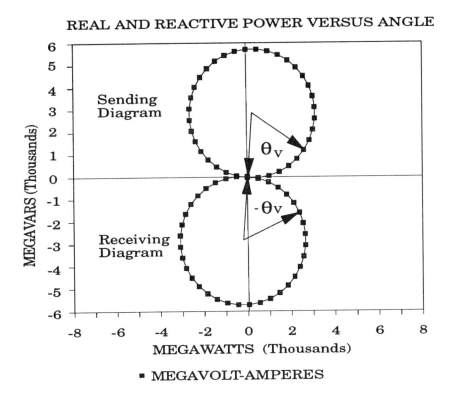

Figure 6-4 Sending and receiving power circle diagrams.

received by the receiving bus. The voltage angle at the point of maximum power transfer occurs at the angle of the series equivalent impedance. This will occur at 90° for a lossless line or at 85.9° for the 345-kV transmission line of our example. We will examine two important relationships regarding transmission line operation. The first is the relationship between reactive power and voltage magnitude. The second is the relationship between real power and voltage angle. A knowledge of the construction of the power circle diagrams will aid in the visualization of these concepts.

124 CHAPTER 6

6.3 THE RELATIONSHIP BETWEEN REACTIVE POWER FLOW AND VOLTAGE MAGNITUDE

The data of Table 6-2 illustrate the relationship between the voltage on the terminals of a typical 345-kV transmission line and the reactive power flow. Note that the voltage angle between the buses is maintained at 0° to separate the effect of the reactive power versus voltage magnitude from the effect of real power versus voltage angle. The line charging which is the reactive power generated by the transmission line is based on the application of the system voltage,

Table 6-2 Reactive Power Flow versus Voltage Magnitude for the Sending Bus

APPARENT POWER (S_S, MVA)	APPARENT POWER ANGLE (θ)	REAL POWER (P_S, MW)	REACTIVE POWER (Q_S, MVAR)	SENDING VOLTAGE (V_S, kV)	RECEIVE VOLTAGE (V_R, kV)
31	-90	0	-31	350	350
22	-88	1	-23	351	350
14	-85	1	-14	352	350
6	-74	2	-6	353	350
3	43	2	2	354	350
11	74	3	11	355	350
19	79	4	19	356	350
28	81	4	27	357	350
36	82	5	36	358	350
45	83	5	44	359	350
53	83	6	53	360	350
62	84	7	61	361	350
71	84	7	70	362	350

POWER CIRCLE DIAGRAMS 125

which is chosen to be 350 kV for this example, to the shunt capacitance. The line charging for this circuit is computed as follows:

$$Q_+ = \frac{V^2}{X_C} = \frac{350^2}{1920} = 63.8 \text{ MVAR} \tag{6-16}$$

Note that the reactive power which is generated by the transmission line flows equally into both terminals, e.g., 31 MVAR into the sending bus and 31 MVAR into the receiving bus, when the voltage on the sending and receiving buses are both maintained at 350 kV.

Table 6-3 Reactive Power Flow versus Voltage Magnitude for the Receiving Bus

APPARENT POWER (S_R, MVA)	APPARENT POWER ANGLE (θ)	REAL POWER (P_R, MW)	REACTIVE POWER (Q_R, MVAR)	SENDING VOLTAGE (V_S, kV)	RECEIVE VOLTAGE (V_R, kV)
31	-90	0	31	350	350
39	-89	-1	39	351	350
47	-89	-1	47	352	350
56	-88	-2	55	353	350
64	-88	-2	64	354	350
72	-88	-3	72	355	350
80	-87	-4	80	356	350
89	-87	-4	89	357	350
97	-87	-5	97	358	350
105	-87	-5	105	359	350
114	-87	-6	114	360	350
122	-87	-7	122	361	350
130	-87	-7	130	362	350

126 CHAPTER 6

The sending bus voltage is then raised in increments of 1 kV to the high voltage limit of 362 kV while the receiving voltage is maintained at 350 kV. Note that the reactive flow from the transmission line to the sending terminal decreases, while the reactive flow from the transmission line to the receiving terminal increases. The conditions at the receiving bus are illustrated in Table 6-3.

6.4 THE RELATIONSHIP BETWEEN REAL POWER FLOW AND VOLTAGE ANGLE

The information in the previous section was based on the relationship between reactive power flow and voltage magnitude. The information contained in this section is based on the relationship between real power flow and voltage angle. In order to separate these relationships,

Table 6-4 Real Power Flow versus Voltage Angle for the Sending Bus

APPARENT POWER (Ss, MVA)	APPARENT POWER ANGLE (θ)	REAL POWER (Ps, MW)	REACTIVE POWER (Qs, MVAR)	VOLTAGE ANGLE (θ_v)
31	90	0	-31	0
61	-34	51	-34	1
108	-20	101	-36	2
157	-14	152	-37	3
206	-11	203	-38	4
256	-8	254	-38	5
307	-7	304	-36	6
357	-6	355	-34	7
407	-4	406	-31	8
457	-3	457	-27	9
508	-3	507	-23	10

POWER CIRCLE DIAGRAMS 127

the voltage magnitude is maintained at 350 kV at both sending and receiving buses, and the voltage angle is adjusted in increments of 1°. These conditions are illustrated in Tables 6-4 and 6-5.

It should be noted that the difference between the real power flow at the sending and receiving buses represents the real power losses that result from the flow of current through the resistance of the phase conductors. Note also that there is no real power transfer between the sending and receiving buses when the voltage angle is 0°.

6.5 SURGE IMPEDANCE LOADING

The characteristic impedance or surge impedance of the transmission line is computed from the methods of Chap 4.

Table 6-5 Real Power Flow versus Voltage Angle for the Receiving Bus

APPARENT POWER (S_R, MVA)	APPARENT POWER ANGLE (θ)	REAL POWER (P_R, MW)	REACTIVE POWER (Q_R, MVAR)	VOLTAGE ANGLE (θ_V)
31	90	0	31	0
57	28	51	27	1
103	12	101	22	2
152	6	152	16	3
202	3	202	9	4
252	0	252	2	5
302	-1	302	-7	6
352	-3	352	-16	7
403	-4	402	-26	8
453	-5	451	-38	9
503	-6	501	-49	10

CHAPTER 6

$$Z_0 = \sqrt{\frac{Z}{Y}} = \sqrt{\frac{R+j\omega L}{G+j\omega C}} \qquad (6\text{-}17)$$

We can approximate the surge impedance by neglecting the series resistance and the shunt conductance. It is interesting to note that the voltage and current are in phase under this condition. This yields the following results for the 345 kV transmission line.

$$\begin{aligned} Z_0 &= \sqrt{\frac{j\omega L}{j\omega C}} \\ &= \sqrt{X_L X_C} \\ &= \sqrt{(42)(1920)} \\ &= 284 \, \Omega \end{aligned} \qquad (6\text{-}18)$$

The surge impedance loading is the level of real power flow at which the reactive power that is absorbed by the flow of current through the series inductance of the transmission line is equal to the reactive power that is produced by the application of the system voltage to the shunt capacitance. The surge impedance loading at 350 kV is

$$P_0 = \frac{V^2}{Z_0} = 431 \text{ MW} \qquad (6\text{-}19)$$

Let us compare this result with the sending and receiving bus data of Tables 6-4 and 6-5. This can best be done by excerpting the key data from the tables which are summarized in Table 6-6.

We can see from the data of Table 6-6 that the surge impedance loading of 431 MW occurs at a voltage angle of between 8° and 9°.

Table 6-6 Surge Impedance Loading Assessment

θ_V (°)	P_S (MW)	P_R (MW)	P_{loss} (MW)	Q_S (MVAR)	Q_R (MVAR)	Q_{loss} (MVAR)
8	406	402	4	-31	-26	-5
9	457	451	6	-27	-38	+11

POWER CIRCLE DIAGRAMS 129

Note from the Q_{loss} column that the transmission line changes from a net producer of 5 MVAR to a net consumer of 11 MVAR. This confirms our point that the surge impedance loading represents a point of real power transfer below which the line charging exceeds the reactive power loss in the line and above which the reactive power loss in the line exceeds the line charging.

6.6 THE POWER ANGLE DIAGRAM

The power-angle diagram provides a graphical illustration of the relationship between real power flow in megawatts and voltage angle. This diagram is based on the data which appear in the above tables for real power flow versus voltage angle on the transmission line.

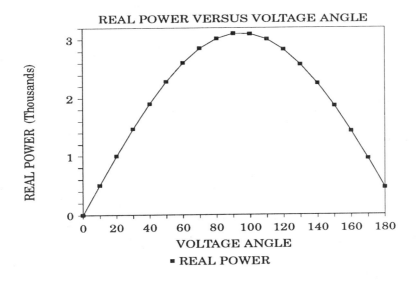

Figure 6-5 Power angle diagram.

Note the reduction in real power transfer for voltage angles beyond 90° which is consistent with our observations from the tables cited above.

CHAPTER 7

SYMMETRICAL COMPONENTS

7.1 INTRODUCTION

The analysis of balanced systems is facilitated by the fact that the magnitudes of the quantities on one phase are equal to the magnitudes of the quantities on the other two phases. Power systems may therefore be represented with a single-line diagram since each phase is identical to the other two. The analysis of one phase may be extended to the other two phases by simply rotating the phase angles by 120° to represent the appropriate phase.

Under unbalanced conditions, however, the magnitudes of the electrical quantities on one phase will differ from those on the other two phases. The analysis of one phase therefore cannot be extended to the other two phases simply by rotating the angles by 120° as in the case of the balanced system.

Dr. Charles L. Fortescue, in a paper published in 1918 entitled "Method of Symmetrical Components Applied to the Solution of Polyphase Networks," developed a method which enabled electric power engineers to represent three unbalanced voltages or currents by positive, negative, and zero sequence quantities. The concepts developed by Dr. Fortescue were advanced by C.F. Wagner and R.D. Evans in their classic book entitled, *Symmetrical Components* which was published in 1933. The following discussion is intended to provide electric power engineers with a basic understanding of the application of symmetrical components to fault analysis.

7.2 FAULT POINT

The diagram of Fig. 7-1 illustrates the location of a fault point on a transmission line within the bulk electric power system.

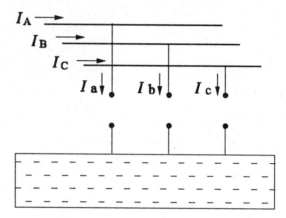

Figure 7-1 System fault point.

The literature regarding symmetrical components often includes the phrase "as viewed from the fault." The application of symmetrical components to fault analysis begins with the following steps:

1. Identify the point in the electric power system where the fault is applied.

2. Reduce the network surrounding the terminals of the fault point.

The following discussions will be based on various line-to-line and line-to-ground faults at this fault point. In order to simplify our analyses, it will be assumed that the impedances of the system surrounding the fault point are the same on the three phases and that no mutual coupling exists between phases. These topics are relegated to more advanced texts on the subject.

SYMMETRICAL COMPONENTS

7.3 POSITIVE SEQUENCE COMPONENTS
The positive sequence voltages and currents are those components which rotate in the positive phase sequence, e.g., *a-b-c*, of the system.

7.3.1 Positive Sequence Current
The positive sequence currents of the three phases are expressed as I_{a1}, I_{b1}, and I_{c1}. These currents are equal in magnitude and separated by angles of 120°. Positive sequence currents are balanced and sum to zero.

$$I_{a1} + I_{b1} + I_{c1} = 0$$

There is therefore no residual current associated with positive sequence currents.

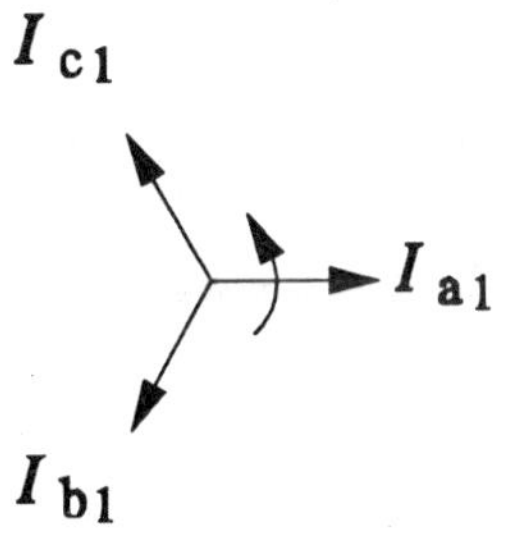

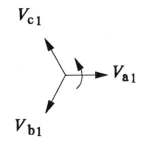

Figure 7-2 Positive sequence currents.

7.3.2 Positive Sequence Voltage
The positive sequence voltages of the three phases are expressed as V_{a1}, V_{b1}, and V_{c1}. These voltages are equal in magnitude and separated by angles of 120°. Positive sequence voltages are balanced and sum to zero.

$$V_{a1} + V_{b1} + V_{c1} = 0$$

There is therefore no residual voltage associated with positive sequence voltages.

Figure 7-3 Positive sequence voltages.

7.3.3 Positive Sequence Impedance

The positive sequence impedances of the three phases are expressed as Z_{a1}, Z_{b1}, and Z_{c1}. These impedances represent the ratio of positive sequence voltage to positive sequence current on each phase. Note that a positive sequence source is included which represents the voltage sources on the electric power system.

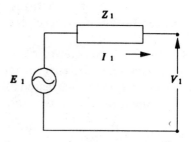

Figure 7-4 Positive sequence network.

The positive sequence impedance of a transmission line can be measured by applying balanced three-phase voltages in a positive phase sequence to the phase conductors and measuring the resultant positive phase sequence currents. The ratio of the positive sequence voltages to the positive sequence currents yields the positive sequence impedance.

7.4 NEGATIVE SEQUENCE COMPONENTS

The negative sequence voltages and currents are those components which rotate in the negative phase sequence, e.g., c-b-a, of the system.

7.4.1 Negative Sequence Current

The negative sequence currents of the three phases are expressed as I_{a2}, I_{b2}, and I_{c2}. These currents are equal in magnitude and separated by angles of 120°. Negative sequence currents are balanced and sum to zero.

$$I_{a2}+I_{b2}+I_{c2}=0$$

There is therefore no residual current associated with negative sequence currents.

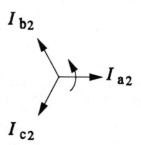

Figure 7-5 Negative sequence currents.

SYMMETRICAL COMPONENTS 135

7.4.2 Negative Sequence Voltages

The negative sequence voltages of the three phases are expressed as V_{a2}, V_{b2} and V_{c2}. These voltages are equal in magnitude and separated by angles of 120°. Negative sequence voltages are balanced and sum to zero.

$$V_{a2}+V_{b2}+V_{c2}=0$$

There is therefore no residual voltage associated with negative sequence voltages.

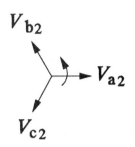

Figure 7-6 Negative sequence voltages.

7.4.3 Negative Sequence Impedance

The negative sequence impedances of the three phases are expressed as Z_{a2}, Z_{b2} and Z_{c2}. These impedances represent the ratio of negative sequence voltage to negative sequence current on each phase. Note that no sources are included in the negative sequence network since all sources on the electric power system rotate in a positive phase sequence.

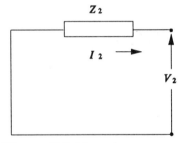

Figure 7-7 Negative sequence network.

The negative sequence impedance of a transmission line can be measured by applying balanced three-phase voltages in a negative phase sequence to the phase conductors and measuring the resultant negative phase sequence currents. The ratio of the negative sequence voltages to the negative sequence currents yields the negative sequence impedance of the transmission line.

It should be noted that positive and negative sequence components are equal for static devices such as transmission lines and transformers but will be different for rotating equipment such as synchronous machines.

7.5 ZERO SEQUENCE COMPONENTS
The zero sequence voltages and currents are those components which rotate in phase.

7.5.1 Zero Sequence Current
The zero sequence currents of the three phases are expressed as I_{a0}, I_{b0}, and I_{c0}. These currents are equal in magnitude and are in phase.

$$I_{a0}+I_{b0}+I_{c0}=3I_0$$

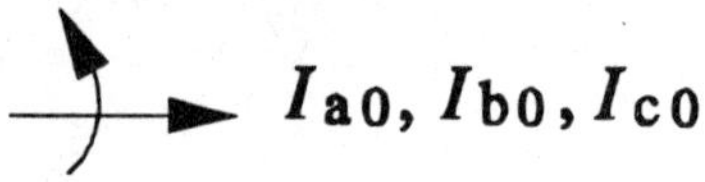

Figure 7-8 Zero sequence currents.

The sum of the zero sequence currents on the three phases is therefore equal to three times the zero sequence current on each of the three phases.

7.5.2 Zero Sequence Voltage
The zero sequence voltages of the three phases are expressed as V_{a0}, V_{b0}, and V_{c0}. These voltages are equal in magnitude and are in phase.

$$V_{a0}+V_{b0}+V_{c0}=3V_0$$

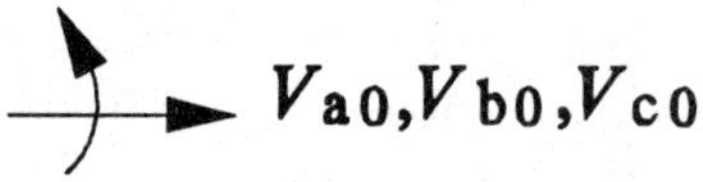

Figure 7-9 Zero sequence voltages.

The sum of the zero sequence voltages on the three phases is therefore equal to three times the zero sequence voltage on each of the three phases.

SYMMETRICAL COMPONENTS

7.5.3 Zero Sequence Impedance

The zero sequence impedances of the three phases are expressed as Z_{a0}, Z_{b0}, and Z_{c0}. These impedances represent the ratio of zero sequence voltage to zero sequence current on each phase. Note that there are no sources included in the zero sequence network since again all of the sources in the electric power system rotate in a positive phase sequence.

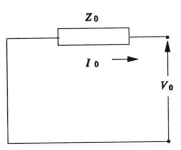

Figure 7-10 Zero sequence network.

The zero sequence impedance of a transmission line can be measured by applying the same voltage to all three phase conductors and measuring the resultant zero phase sequence currents. The ratio of the zero sequence voltages to the zero sequence currents yields the zero sequence impedance of the transmission line.

The analysis of unbalanced faults is simplified by the consideration that the positive and negative sequence components of voltage and current are balanced and sum to zero. There is therefore no return path required for positive or negative sequence components of fault current. The zero sequence voltages and currents, however, do not sum to zero. The three zero sequence currents which flow on the transmission line due to unbalanced faults are in phase and must have a return path through the grounded neutrals of the power system. It should be noted that zero sequence currents cannot flow through delta or ungrounded wye connections since no return paths exist for these cases.

7.6 PHASE QUANTITIES VERSUS SEQUENCE QUANTITIES

The phase voltages and currents can be expressed as the sum of the positive, negative, and zero sequence voltages and currents.

$$V_a = V_{a1} + V_{a2} + V_{a0} \quad V_b = V_{b1} + V_{b2} + V_{b0} \quad V_c = V_{c1} + V_{c2} + V_{c0}$$

$$I_a = I_{a1} + I_{a2} + I_{a0} \quad I_b = I_{b1} + I_{b2} + I_{b0} \quad I_c = I_{c1} + I_{c2} + I_{c0}$$

The sequence voltages and currents may also be expressed in terms of the phase voltages and currents.

CHAPTER 7

$$V_1 = \frac{V_a + aV_b + a^2 V_c}{3} \qquad V_2 = \frac{V_a + a^2 V_b + aV_c}{3} \qquad V_0 = \frac{V_a + V_b + V_c}{3}$$

$$I_1 = \frac{I_a + aI_b + a^2 I_c}{3} \qquad I_2 = \frac{I_a + a^2 I_b + aI_c}{3} \qquad I_0 = \frac{I_a + I_b + I_c}{3}$$

The operator a is a unit vector which has a magnitude equal to 1 and an angle equal to 120°.

$$a = 1\angle 120°$$

The sequence voltages V_1, V_2, V_0 and the sequence currents I_1, I_2, I_0 are assumed to represent the a-phase quantities. The corresponding quantities for b-phase and c-phase may be found by application of the unit vector a.

Table 7-1 Sequence Voltages and Currents

PHASE	POSITIVE SEQUENCE COMPONENTS	NEGATIVE SEQUENCE COMPONENTS	ZERO SEQUENCE COMPONENTS
a	$I_{a1} = I_1$ $V_{a1} = V_1$	$I_{a2} = I_2$ $V_{a2} = V_2$	$I_{a0} = I_0$ $V_{a0} = V_0$
b	$I_{b1} = a^2 I_1$ $V_{b1} = a^2 V_1$	$I_{b2} = aI_2$ $V_{b2} = aV_2$	$I_{b0} = I_0$ $V_{b0} = V_0$
c	$I_{c1} = aI_1$ $V_{c1} = aV_1$	$I_{c2} = a^2 I_2$ $V_{c2} = a^2 V_2$	$I_{c0} = I_0$ $V_{c0} = V_0$

7.7 LINE-TO-LINE FAULT

Consider the diagram of Fig. 7-11 which depicts the line-to-line fault condition.

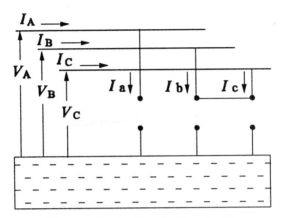

Figure 7-11 Line-to-line fault.

The following relationships again describe the positive, negative, and zero sequence currents as functions of the three fault currents.

$$I_1 = \frac{I_a + aI_b + a^2 I_c}{3}$$

$$I_2 = \frac{I_a + a^2 I_b + aI_c}{3}$$

$$I_0 = \frac{I_a + I_b + I_c}{3}$$

The following information can be determined from the diagram for the line-to-line fault:

$$I_b = -I_c$$
$$I_a = 0$$
$$I_1 = \frac{aI_b + a^2 I_c}{3}$$
$$I_2 = \frac{a^2 I_b + aI_c}{3}$$
$$I_0 = \frac{I_b + I_c}{3}$$

It is important to note that since $I_b = -I_c$, then $I_0 = 0$. There is therefore no zero sequence current in a phase fault. The following relationships may therefore be established for the sequence connection diagram for the line-to-line fault.

$$I_b = I$$
$$I_c = -I_b = -I$$

The sequence currents may now be computed as follows:

$$I_1 = \frac{aI - a^2 I}{3} = \frac{(a-a^2)I}{3}$$
$$I_2 = \frac{a^2 I - aI}{3} = -\frac{(a-a^2)I}{3}$$
$$I_1 = -I_2$$
$$I_0 = \frac{I-I}{3} = 0$$

Note that the positive and negative sequence networks are connected in opposite polarity since $I_1 = -I_2$, and the zero sequence network is open-circuited since $I_0 = 0$.

SYMMETRICAL COMPONENTS

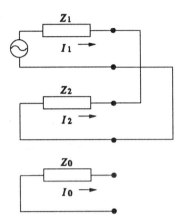

Figure 7-12 Line-to-line fault sequence network.

The phase relationships for the line-to-line fault may be derived on the basis of the current conditions as viewed from the fault. We have at our disposal the following information:

1. The a-phase current as viewed from the fault is equal to zero.

$$I_a = 0$$

2. The b-phase current as viewed from the fault is equal in magnitude and opposite in direction to the c-phase current.

$$I_b = -I_c$$

3. There is no zero sequence component of current in the line-to-line fault.

$$I_{a0} = I_{b0} = I_{c0} = 0$$

We are now equipped to determine the correct phase relationships for the sequence currents. We will first assume a reference for the positive sequence currents as shown below.

CHAPTER 7

I_{c1} I_{a1} I_{b1}

We will then orient the negative sequence current vectors to cancel the positive sequence current for a-phase which we know to be zero.

I_{c2} I_{a2} I_{b2}

We know that there are no zero sequence components of current. We may now reconstruct the phase currents from their symmetrical components.

A-PHASE CURRENT	B-PHASE CURRENT	C-PHASE CURRENT	RESULT
$I_{a2} \longleftarrow \cdot \longrightarrow I_{a1}$	I_{b1}, I_{b2}, I_b	I_{c2}, I_{c1}, I_c	I_c, I_b

Table 7-2 is a summary of the phase relationships for positive, negative, and zero sequence currents for the various line-to-line fault conditions.

SYMMETRICAL COMPONENTS 143

Table 7-2 Sequence Currents for Line-to-Line Faults

FAULT CATEGORY	POSITIVE SEQUENCE CURRENTS	NEGATIVE SEQUENCE CURRENTS	ZERO SEQUENCE CURRENTS
a-b	I_{c1}, I_{a1}, I_{b1}	I_{a2}, I_{b2}, I_{c2}	—
b-c	I_{c1}, I_{a1}, I_{b1}	I_{c2}, I_{a2}, I_{b2}	—
c-a	I_{c1}, I_{a1}, I_{b1}	I_{b2}, I_{c2}, I_{a2}	—

The relationships for the sequence voltages may be developed on the basis that the voltage on the unfaulted phase V_a is relatively unaffected and the faulted phase voltages to ground are equal, e.g., $V_b = V_c$ at the fault point. It will also be assumed that the positive, negative, and zero sequence networks are purely inductive, and the voltage phasors will therefore lead the current phasors by 90°.

We are now equipped to determine the correct phase relationships for the sequence voltages. We will first assume a reference for the positive sequence voltages as shown below.

144 CHAPTER 7

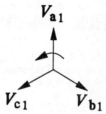

We will then orient the negative sequence voltage vectors to sum with the positive sequence voltage for a-phase which we know is the "healthy" phase.

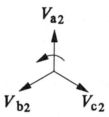

The zero sequence voltage is the product of the zero sequence current and the zero sequence impedance.

$$V_0 = I_0 Z_0$$

It was shown earlier that there is no zero sequence component of current in the phase-to-phase fault, and hence there is no zero sequence component of voltage. We may now reconstruct the phase voltages from their symmetrical components.

A-PHASE VOLTAGE	B-PHASE VOLTAGE	C-PHASE VOLTAGE	RESULT
V_{a2} V_{a1}	V_b V_{b1} V_{b2}	V_{c1} V_{c2} V_c	V_a $V_b = V_c$

Table 7-3 is a summary of the phase relationships for positive, negative, and zero sequence voltages for the various line-to-line fault conditions.

Table 7-3 Sequence Voltages for Line-to-Line Faults

FAULT CATEGORY	POSITIVE SEQUENCE VOLTAGES	NEGATIVE SEQUENCE VOLTAGES	ZERO SEQUENCE VOLTAGES
a-b	V_{a1}, V_{c1}, V_{b1}	V_{b2}, V_{c2}, V_{a2}	—
b-c	V_{a1}, V_{c1}, V_{b1}	V_{a2}, V_{b2}, V_{c2}	—
c-a	V_{a1}, V_{c1}, V_{b1}	V_{c2}, V_{a2}, V_{b2}	—

The phase-to-phase voltages and the phase currents for an actual line-to-line fault are shown in Fig. 7-13. Note the increase in the b-phase and c-phase currents from load levels to fault levels. The two currents are opposite in polarity as we have predicted in our analysis.

Note also that the fault voltage V_{bc} is relatively collapsed in comparison to the V_{ab} and V_{ca} voltages. The fault voltage would appear to be collapsed to a greater extent if the fault occurred closer to the measuring terminal.

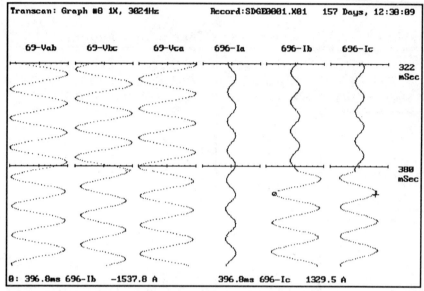

Figure 7-13 Phase voltages and currents for a line-to-line fault. (*Courtesy of Mehta Tech, Inc.*)

7.8 LINE-TO-GROUND FAULT

Consider the diagram of Fig. 7-14 which depicts a line-to-ground fault condition on A-phase.

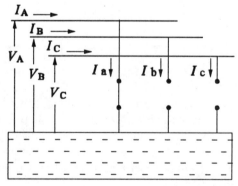

Figure 7-14 Line-to-ground fault.

SYMMETRICAL COMPONENTS 147

The relationships between the positive, negative, and zero sequence currents and the three fault currents may be stated as follows:

$$I_1 = \frac{I_a + aI_b + a^2 I_c}{3}$$

$$I_2 = \frac{I_a + a^2 I_b + aI_c}{3}$$

$$I_0 = \frac{I_a + I_b + I_c}{3}$$

The following information can be determined from the diagram for the line-to-ground fault:

$$I_b = I_c = 0$$

$$I_1 = I_2 = I_0 = \frac{I_a}{3}$$

These relationships form the basis for the connections of sequence diagrams which can be used to compute fault currents. Since the same current flows through the positive, negative, and zero sequence networks, then the sequence networks can be constructed accordingly.

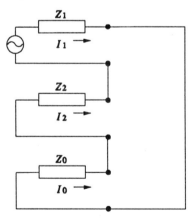

Figure 7-15 Line-to-ground sequence network.

The phase relationships for the line-to-ground fault may be derived on the basis of the current conditions as viewed from the fault. We have

148 **CHAPTER 7**

at our disposal the following information:

1. The same current flows through the positive, negative, and zero sequence network.

$$I_1 = I_2 = I_0$$

2. The symmetrical components for the b-phase and c-phase currents sum to zero.

$$I_b = I_{b1} + I_{b2} + I_{b0} = 0$$
$$I_c = I_{c1} + I_{c2} + I_{c0} = 0$$

We are now equipped to determine the correct phase relationships for the sequence currents. We will again assume the following reference for the positive sequence currents.

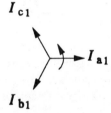

We will then orient the negative sequence current so that the a-phase positive sequence current is equal to the negative sequence a-phase current in magnitude and phase.

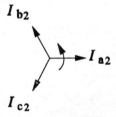

We orient the zero sequence current so that the a-phase zero sequence current is also equal in magnitude and phase to the positive and negative sequence components of the a-phase current.

SYMMETRICAL COMPONENTS 149

↗ I_{a0}, I_{b0}, I_{c0}

We may now reconstruct the phase currents for the line-to-ground fault from their symmetrical components.

A-PHASE CURRENT	B-PHASE CURRENT	C-PHASE CURRENT	RESULT
I_{a1} I_{a2} I_{a0}	I_{b0}, I_{b2}, I_{b1}	I_{c2}, I_{c1}, I_{c0}	I_a

Table 7-4 is a summary of the phase relationships for positive, negative, and zero sequence currents and the associated phase currents for the various line-to-ground fault conditions.

Table 7-4 Sequence Currents for Line-to-Ground Faults

FAULT CATEGORY	POSITIVE SEQUENCE CURRENTS	NEGATIVE SEQUENCE CURRENTS	ZERO SEQUENCE CURRENTS
a-g	I_{c1}, I_{a1}, I_{b1}	I_{b2}, I_{a2}, I_{c2}	I_{a0}, I_{b0}, I_{c0}
b-g	I_{c1}, I_{a1}, I_{b1}	I_{a2}, I_{c2}, I_{b2}	I_{a0}, I_{b0}, I_{c0}
c-g	I_{c1}, I_{a1}, I_{b1}	I_{c2}, I_{b2}, I_{a2}	I_{a0}, I_{b0}, I_{c0}

The voltage relationships for the line-to-ground fault may be derived on the basis of the voltage conditions as viewed from the fault. Note that the a-phase voltage is zero at the fault point so that the positive, negative, and zero sequence components of voltage for that phase must sum to zero.

$$V_a = V_{a1} + V_{a2} + V_{a0} = 0$$

We are now equipped to determine the correct phase relationships for the sequence voltages. We will again assume the following reference for the positive sequence voltages.

SYMMETRICAL COMPONENTS

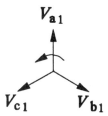

We will then orient the negative and zero sequence voltages to cancel the positive sequence voltage on a-phase so that V_a is equal to zero.

We may now reconstruct the phase voltages for the line-to-ground fault from their symmetrical components.

A-PHASE VOLTAGE	B-PHASE VOLTAGE	C-PHASE VOLTAGE	RESULT
V_{a1} ↑ V_{a2} ↓ V_{a0} ↓	V_b, V_{b0}, V_{b1}, V_{b2}	V_c, V_{c0}, V_{c2}, V_{c1}	V_b V_c

Table 7-5 is a summary of the phase relationships for positive, negative, and zero sequence voltages for the various line-to-ground fault conditions.

CHAPTER 7

Table 7-5 Sequence Voltages for Line-to-Ground Faults

FAULT CATEGORY	POSITIVE SEQUENCE VOLTAGES	NEGATIVE SEQUENCE VOLTAGES	ZERO SEQUENCE VOLTAGES
a-g	V_{a1}, V_{c1}, V_{b1}	V_{c2}, V_{b2}, V_{a2}	V_{a0}, V_{b0}, V_{c0}
b-g	V_{a1}, V_{c1}, V_{b1}	V_{b2}, V_{a2}, V_{c2}	V_{a0}, V_{b0}, V_{c0}
c-g	V_{a1}, V_{c1}, V_{b1}	V_{a2}, V_{c2}, V_{b2}	V_{a0}, V_{b0}, V_{c0}

A digital fault recording of an actual a-phase-to-ground fault is illustrated in Fig. 7-16. Note the collapse of the a-phase voltage and the increase of the a-phase and neutral currents. This is a relatively close-in fault since the a-phase voltage is nearly completely collapsed.

SYMMETRICAL COMPONENTS 153

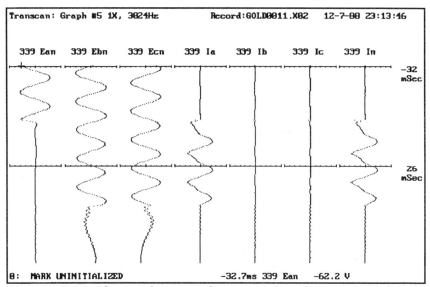

Figures 7-16 Phase voltages and currents for a line-to-ground fault. (*Courtesy of Mehta Tech, Inc.*)

7.9 DOUBLE-LINE-TO-GROUND FAULT

Consider the diagram of Fig. 7-17 which depicts the double-line-to-ground fault condition.

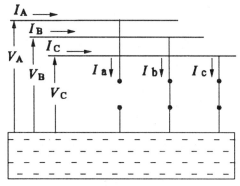

Figure 7-17 Double-line-to-ground fault.

CHAPTER 7

The following relationships again describe the positive, negative, and zero sequence currents as functions of the three fault currents.

$$I_1 = \frac{I_a + aI_b + a^2 I_c}{3}$$

$$I_2 = \frac{I_a + a^2 I_b + aI_c}{3}$$

$$I_0 = \frac{I_a + I_b + I_c}{3}$$

The following information can be determined from the diagram for the double-line-to-ground fault:

$$I_a = 0$$

$$I_1 = \frac{aI_b + a^2 I_c}{3}$$

$$I_2 = \frac{a^2 I_b + aI_c}{3}$$

$$I_0 = \frac{I_b + I_c}{3}$$

Note that the following relationships may be established for the sequence connection diagram for the double-line-to-ground fault.

$$I_0 = \frac{I_b + I_c}{3}$$

$$I_1 + I_2 = \frac{aI_b + a^2 I_c + a^2 I_b + aI_c}{3}$$

$$= \frac{(a + a^2)I_b + (a + a^2)I_c}{3}$$

$$= \frac{-I_b - I_c}{3}$$

$$= -I_0$$

The positive and negative sequence currents are summed to form the negative of the zero sequence current since $I_1 + I_2 = -I_0$.

SYMMETRICAL COMPONENTS 155

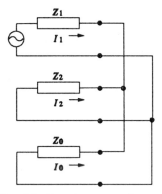

Figure 7-18 Double-line-to-ground sequence network.

The phase relationships for the double-line-to-ground fault may be derived on the basis of the current conditions as viewed from the fault. We have at our disposal the following information:

1. The a-phase current as viewed from the fault is equal to zero.

$$I_a = 0$$

2. There is a zero sequence component of current in the double-line-to-ground fault since the three phase currents do not sum to zero.

$$I_0 = \frac{I_a + I_b + I_c}{3}$$
$$I_0 \neq 0$$

We are now equipped to determine the correct phase relationships for the sequence currents. We will assume a reference for the positive sequence currents as in the previous examples.

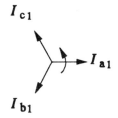

CHAPTER 7

We will then orient the negative and zero sequence current vectors to cancel the positive sequence current for a-phase which we know to be zero.

$$I_{a2} \quad I_{c2} \quad I_{b2} \qquad I_{a0}, I_{b0}, I_{c0}$$

We may now reconstruct the phase currents from their symmetrical components.

A-PHASE CURRENT	B-PHASE CURRENT	C-PHASE CURRENT	RESULT
I_{a0} I_{a2} I_{a1}	I_b, I_{b1}, I_{b2}, I_{b0}	I_{c0}, I_c, I_{c2}, I_{c1}	I_c, I_b

Table 7-6 is a summary of the phase relationships for positive, negative, and zero sequence currents for the various line-to-line fault conditions.

SYMMETRICAL COMPONENTS

Table 7-6 Sequence Currents for Double-Line-to-Ground Faults

FAULT CATEGORY	POSITIVE SEQUENCE CURRENTS	NEGATIVE SEQUENCE CURRENTS	ZERO SEQUENCE CURRENTS
a-b-g	I_{c1}, I_{a1}, I_{b1}	I_{a2}, I_{b2}, I_{c2}	I_{a0}, I_{b0}, I_{c0}
b-c-g	I_{c1}, I_{a1}, I_{b1}	I_{c2}, I_{a2}, I_{b2}	I_{a0}, I_{b0}, I_{c0} ←
c-a-g	I_{c1}, I_{a1}, I_{b1}	I_{b2}, I_{c2}, I_{a2}	I_{a0}, I_{b0}, I_{c0}

The voltage relationships for the double-line-to-ground fault may be derived on the basis of the voltage conditions as viewed from the fault. We have at our disposal the following information:

1. The a-phase voltage V_a is relatively unaffected, and the b-phase and c-phase voltages are equal to zero at the fault point.

$$V_b = V_c = 0$$

2. There is a zero sequence component of voltage in the double-line-to-ground fault since the three phase voltages do not sum to zero.

158   CHAPTER 7

$$V_0 = \frac{V_a + V_b + V_c}{3}$$

$$= \frac{V_a}{3}$$

We are now equipped to determine the correct phase relationships for the sequence voltages. We will assume a reference for the positive sequence voltages as in the previous examples.

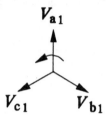

We will then orient the negative and zero sequence vectors to sum with the positive sequence voltage for a-phase which we know to be the "healthy" phase.

We may now reconstruct the phase voltages from their symmetrical components.

A-PHASE VOLTAGE	B-PHASE VOLTAGE	C-PHASE VOLTAGE	RESULT
V_{a0} V_{a2} V_{a1}	V_{b0} V_{b1} V_{b2}	V_{c1} V_{c0} V_{c2}	V_a

SYMMETRICAL COMPONENTS 159

Table 7-7 is a summary of the phase relationships for positive, negative and zero sequence voltages for the various double-line-to-ground fault conditions.

Table 7-7 Sequence Voltages for Double-Line-to-Ground Faults

FAULT CATEGORY	POSITIVE SEQUENCE VOLTAGE	NEGATIVE SEQUENCE VOLTAGE	ZERO SEQUENCE VOLTAGE
a-b-g	V_{a1}, V_{c1}, V_{b1}	V_{b2}, V_{c2}, V_{a2}	V_{a0}, V_{b0}, V_{c0}
b-c-g	V_{a1}, V_{c1}, V_{b1}	V_{a2}, V_{b2}, V_{c2}	V_{a0}, V_{b0}, V_{c0}
c-a-g	V_{a1}, V_{c1}, V_{b1}	V_{c2}, V_{a2}, V_{b2}	V_{a0}, V_{b0}, V_{c0}

7.10 APPLICATIONS TO SYSTEM PROTECTION

Ground faults are characterized by positive, negative, and zero sequence components. Phase faults are characterized by only positive and negative sequence components. Zero sequence components which are computed from the sum of the three phase currents are not present in phase faults since one current at the fault point is zero and the other two are equal and opposite in polarity.

Negative sequence fault detectors are devices which are used for the detection of both phase and ground faults since they are present in both types of faults. Negative sequence voltages and currents appear as the result of phase imbalances during steady state operation or to a much greater extent as the result of severe imbalances due to faults on the system.

Zero sequence fault detectors are devices which are used for the detection of ground faults. Zero sequence voltages and currents can be measured directly by summing the outputs of voltage and current transformers that are connected to the three phases of the system. Zero sequence current can also be measured as the residual current through the neutral of a grounded-wye transformer bank. Positive and negative sequence components sum to zero and the zero sequence or residual component that remains is measured.

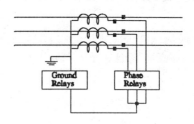

Figure 7-19 Protective relays.

This fact can be used to provide discrimination between phase and ground faults since phase fault currents include positive and negative sequence components but no zero sequence components. An overcurrent relay which operates on zero sequence current may be included in the ground fault relays as a ground fault detector (GFD). The operating coil of the overcurrent relay receives the sum of the three phase currents from the residual connection of the current transformers shown in Fig. 7-19. Operating current flows through the circuit when the sum of the three phase currents is not equal to zero. This condition would signify a ground fault condition, and the operating contact of the overcurrent relay would close to arm the ground fault protection scheme.

A phase fault detector (PFD) may consist of an overcurrent relay which is set above the emergency rating of a circuit to provide discrimination between overload conditions and phase fault conditions. The positive and negative sequence components which comprise phase fault currents are balanced and sum to zero. The zero sequence component is not present in a phase fault and is therefore equal to zero. The current transformers will therefore not produce operating current through the coil of the ground fault detector for the phase fault condition.

CHAPTER 8

TRANSIENT ANALYSIS

The network response of an electric power system consists of a forced response that is due to the presence of generating sources and a natural response that is due to the energy stored in the inductance and capacitance of the system and dissipated in the resistance. The rotor of a large three-phase, sixty-pole hydroelectric generator, for example, may weigh 600 tons and rotate at a synchronous speed of 120 revolutions per minute (rpm) to produce 60-Hz voltages and currents. The response of the system to voltages and currents produced by the generator is referred to as the forced response.

The natural response may be visualized by the removal of the generating sources from the system. Energy is stored in the magnetic fields that result from the flow of current through the inductance of the power system conductors and in the electric fields that result from the application of the operating voltage to the system capacitance. The transfer of energy between the electric and magnetic fields occurs at the natural frequency of the system which is based on the values of inductance, capacitance, and resistance. This transfer of energy will occur until all of the energy initially stored in the circuit is dissipated by the flow of current through the resistance of the phase conductors. We will begin our analysis of the natural response with a review of the various damping conditions associated with power system transients. An integrodifferential equation is one in which both differential and integral terms are present. An example of such an equation is the following:

$$L\frac{di}{dt} + Ri + \frac{1}{C}\int idt = V_0 \qquad (8\text{-}1)$$

This equation is based on Kirchoff's voltage law for a series circuit consisting of an inductor, a resistor, and a capacitor with an initial voltage V_0 impressed across the terminals of the capacitor.

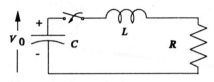

Figure 8-1 *RLC* circuit.

Differentiation of this equation yields

$$L\frac{d^2i}{dt^2}+R\frac{di}{dt}+\frac{1}{C}i=0 \qquad (8\text{-}2)$$

The equation may be rewritten as follows:

$$\frac{d^2i}{dt^2}+\frac{R}{L}\frac{di}{dt}+\frac{1}{LC}i=0 \qquad (8\text{-}3)$$

Transforming variables from the time domain to the complex frequency domain yields the following characteristic equation:

$$s^2+\frac{R}{L}s+\frac{1}{LC}=0 \qquad (8\text{-}4)$$

The roots of the equation may be found from the quadratic formula:

$$s_1, s_2 = -b \pm \frac{\sqrt{b^2-4ac}}{2a} \qquad (8\text{-}5)$$

Substituting coefficients yields

$$a=1 \qquad b=\frac{R}{L} \qquad c=\frac{1}{LC} \qquad (8\text{-}6a,b,c)$$

The quadratic equation with the substitution of coefficients yields

$$s_1, s_2 = -\frac{R}{2L} \pm \sqrt{\left(\frac{R}{2L}\right)^2 - \frac{1}{LC}} \qquad (8\text{-}7)$$

The general solution may be written as

$$i = C_1 \varepsilon^{s_1 t} + C_2 \varepsilon^{s_2 t} \qquad (8\text{-}8)$$

TRANSIENT ANALYSIS

We will now consider the various cases for the roots of the characteristic equation.

8.1 UNDAMPED CASE

The case in which $R=0$, e.g., the undamped or lossless case, yields the following roots:

$$s_1, s_2 = \pm\sqrt{-\frac{1}{LC}} \qquad (8\text{-}9)$$

We now introduce the undamped natural frequency of the circuit which is expressed as follows:

$$\omega_0 = \sqrt{\frac{1}{LC}} \qquad (8\text{-}10)$$

The roots of the characteristic equation may now be written as follows:

$$\begin{aligned} s_1, s_2 &= \pm\sqrt{-\omega_0^2} \\ &= \pm j\omega_0 \end{aligned} \qquad (8\text{-}11)$$

A plot of the characteristic roots on the s plane is illustrated in Fig. 8-2.

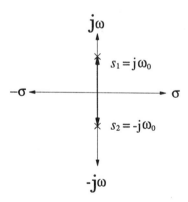

Figure 8-2 Characterisic roots for the undamped case.

CHAPTER 8

The roots are found to be purely imaginary. The general solution may be written as

$$i = C_1 \varepsilon^{j\omega_0 t} + C_2 \varepsilon^{-j\omega_0 t} \quad (8\text{-}12)$$

The particular solution is obtained by evaluating the initial conditions at $t=0$.

$$i(0) = C_1 + C_2 = 0 \quad (8\text{-}13)$$

$$C_1 = -C_2 \quad (8\text{-}14)$$

Let $C_1 = -C_2 = C_3$. Then $C_1 = C_3$ and $C_2 = -C_3$. The particular solution is then

$$\begin{aligned} i &= C_3 \varepsilon^{j\omega_0 t} - C_3 \varepsilon^{-j\omega_0 t} \\ &= C_3 (\varepsilon^{j\omega_0 t} - \varepsilon^{-j\omega_0 t}) \end{aligned} \quad (8\text{-}15)$$

The total current therefore consists of a current phasor of magnitude C_3 rotating in a positive direction with radian frequency ω_0 and a current phasor of magnitude $-C_3$ rotating in a negative direction with radian frequency $-\omega_0$. The two current phasors are shown in Fig. 8-3.

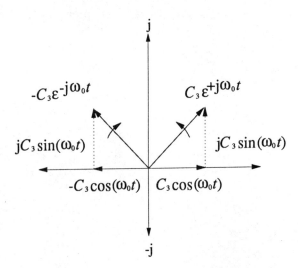

Figure 8-3 Current phasors for the undamped case.

TRANSIENT ANALYSIS 165

The particular solution may be resolved by invoking Euler's identity.

$$\varepsilon^{j\omega_0 t} = \cos(\omega_0 t) + j\sin(\omega_0 t) \quad (8\text{-}16)$$

$$\varepsilon^{-j\omega_0 t} = \cos(\omega_0 t) - j\sin(\omega_0 t) \quad (8\text{-}17)$$

Computing the difference of the terms yields

$$\varepsilon^{j\omega_0 t} - \varepsilon^{-j\omega_0 t} = \cos(\omega_0 t) + j\sin(\omega_0 t) - \cos(\omega_0 t) + j\sin(\omega_0 t)$$
$$= 2j\sin(\omega_0 t) \quad (8\text{-}18)$$
$$= 2\sin\left(\omega_0 t + \frac{\pi}{2}\right)$$

The addition of the two current phasors is illustrated graphically in Fig. 8-4. The real components $C_3\cos(\omega_0 t)$ and $-C_3\cos(\omega_0 t)$ are equal in magnitude and opposite in phase and the imaginary components $C_3\sin(\omega_0 t)$ are equal in magnitude and are in phase.

The particular solution may be rewritten as a single sinusoid with a magnitude $I_m = 2C_3$, a radian frequency of ω_0, and a phase angle of 90°.

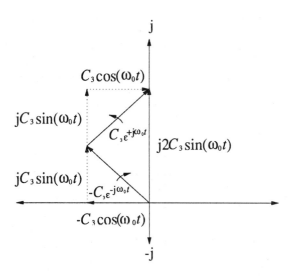

Figure 8-4 Sum of phasors for the undamped case.

$$i = I_m \sin\left(\omega_0 t + \frac{\pi}{2}\right) \qquad (8\text{-}19)$$

The time domain representation for the undamped case is illustrated in Fig. 8-5. Note that the total current consists only of an oscillatory component and that the oscillations will continue indefinitely due to the absence of losses.

The characteristic roots for the undamped case are summarized in Table 8-1.

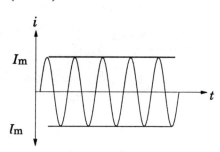

Figure 8-5 Time domain plot for the underdamped case.

Table 8-1 Characteristic Roots for the Undamped Case

ROOTS	σ	jω
s_1	0	ω_0
s_2	0	$-\omega_0$

8.2 CRITICALLY DAMPED CASE
The critically damped case is achieved where the radical term is equal to zero.

$$\left(\frac{R}{2L}\right)^2 - \frac{1}{LC} = 0 \qquad (8\text{-}20)$$

The value of the resistance R which yields this condition is referred to as the critical resistance R_c. This value is obtained as follows:

TRANSIENT ANALYSIS

$$R = 2\sqrt{\frac{L}{C}} \tag{8-21}$$

The damping ratio ζ is the ratio of the actual circuit resistance R to the critical resistance R_C. This term is expressed as follows:

$$\zeta = \frac{R}{R_C} = \frac{R}{2\sqrt{L/C}} = \frac{R}{2}\sqrt{\frac{C}{L}} \tag{8-22}$$

The damping ratio for the critically damped case in which $R = R_C$ is expressed as follows:

$$\zeta = \frac{R}{R_C} = 1 \tag{8-23}$$

The damping ratio for the undamped case cited previously in which $R = 0$ is expressed as follows:

$$\zeta = \frac{R}{R_C} = 0 \tag{8-24}$$

The critically damped condition yields the following roots:

$$s_1, s_2 = -\frac{R}{2L} \tag{8-25}$$

Note that there is no oscillatory component in the critically damped case.

The term $\pi/2$ may be expressed in terms of the undamped natural frequency ω_0 and the damping ratio ζ as follows:

$$\zeta = \frac{R}{2}\sqrt{\frac{C}{L}} \tag{8-26}$$

$$\omega_0 = \sqrt{\frac{1}{LC}} \tag{8-27}$$

CHAPTER 8

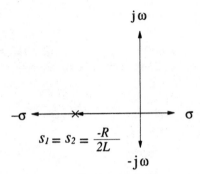

Figure 8-6 Characteristic roots for the critically damped case.

The roots are therefore found to be real and repeated. The general solution may be written as

$$i = C_1 \varepsilon^{-(R/2L)t} + C_2 t \varepsilon^{-(R/2L)t} \tag{8-28}$$

The total current consists of a current magnitude C_1 which decays exponentially at a rate of $-R/2L$ and a current magnitude C_2 which decays at the same rate and which is scaled by the independent variable t. The general solution may therefore be rewritten as follows:

$$i = C_1 \varepsilon^{-\omega_0 \zeta t} + C_2 t \varepsilon^{-\omega_0 \zeta t} \tag{8-29}$$

The time domain representation for the critically damped case is illustrated in Fig. 8-7. Note that all of the initial energy stored in the capacitor is dissipated in the resistor during the initial discharge of current and that no recharging of the capacitor occurs.

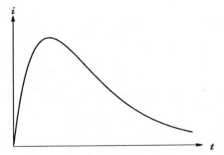

Figure 8-7 Time domain representation for the critically damped case.

TRANSIENT ANALYSIS 169

The critically damped case represents the minimum amount of resistance that is necessary to eliminate the oscillatory component. The characteristic roots for the critically damped case are summarized in Table 8-2.

Table 8-2 Characteristic Roots for the Critically Damped Case

ROOTS	σ	$j\omega$
s_1	$-\omega_0\zeta$	0
s_2	$-\omega_0\zeta$	0

8.3 UNDERDAMPED CASE

The underdamped case is the case in which the actual circuit resistance is less than the critical resistance. The damping ratio for this case is expressed by the following inequality:

$$0 > \zeta > 1 \tag{8-30}$$

The general expression for the roots of the characteristic equation is restated in terms of the undamped natural frequency and the damping ratio as follows:

$$\begin{aligned} s_1, s_2 &= -\omega_0\zeta \pm \sqrt{(\omega_0\zeta)^2 - \omega_0^2} \\ &= -\omega_0\zeta \pm \sqrt{\omega_0^2(\zeta^2 - 1)} \\ &= -\omega_0\zeta \pm \sqrt{(-\omega_0^2)(1 - \zeta^2)} \\ &= -\omega_0\zeta \pm j\omega_0\sqrt{1 - \zeta^2} \end{aligned} \tag{8-31}$$

The S-plane representation of the characteristic roots for the underdamped case is illustrated in Fig. 8-8.

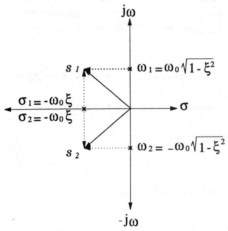

Figure 8-8 Characteristic roots for the underdamped case.

The time domain representation for the underdamped case is illustrated in Fig. 8-9. Note that the total current consists of an oscillatory component and a real component and that the oscillations decay in magnitude due to the dissipation of energy in the resistor.

The characteristic roots for the underdamped case are complex conjugates and are summarized in Table 8-3.

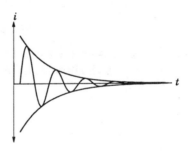

Figure 8-9 Time domain plot for the underdamped case.

TRANSIENT ANALYSIS

Table 8-3 Characteristic Roots for the Underdamped Case

ROOTS	σ	$j\omega$
s_1	$-\omega_0\zeta$	$\omega_0\sqrt{1-\zeta^2}$
s_2	$-\omega_0\zeta$	$-\omega_0\sqrt{1-\zeta^2}$

Let us find the particular solution for the underdamped condition. Equation (8-1) is restated for convenience as follows:

$$L\frac{di}{dt}+Ri+\frac{1}{C}\int idt = V_0 \qquad (8\text{-}32)$$

Dividing through by the inductance yields

$$\frac{di}{dt}+\frac{R}{L}i+\frac{1}{LC}\int idt = \frac{V_0}{L} \qquad (8\text{-}33)$$

We will now establish the following initial condition of current:

$$i(0)=0 \qquad (8\text{-}34)$$

This gives rise to the following initial condition for the first derivative of the current:

$$\frac{di(t)}{dt}=\frac{V_0}{L} \qquad (8\text{-}35)$$

The general solution was found to be

$$i = C_1\varepsilon^{s_1 t}+C_2\varepsilon^{s_2 t} \qquad (8\text{-}36)$$

Conditions at $i(0)=0$ yield

$$i = C_1+C_2 = 0 \qquad (8\text{-}37)$$

Letting $C_1 = k$ and $C_2 = -k$ yields the following:

$$i(t)=k(\varepsilon^{s_1 t}-\varepsilon^{s_2 t}) \qquad (8\text{-}38)$$

The first derivative of the current may then be expressed as follows:

$$\frac{di(t)}{dt}=k(s_1\varepsilon^{s_1 t}-s_2\varepsilon^{s_2 t}) \qquad (8\text{-}39)$$

Evaluating the first derivative of current at $t=0$ and equating this expression with the initial conditions found in (8-35) yields

$$\frac{di(0)}{dt}=ks_1-ks_2=\frac{V_0}{L} \qquad (8\text{-}40)$$

The characteristic roots for the underdamped case were found to be of the form:

$$s_1,\ s_2 = -\sigma \pm j\omega \qquad (8\text{-}41)$$

The initial conditions for the first derivative may now be used to solve for k as follows:

$$k(s_1-s_2)=\frac{V_0}{L} \qquad (8\text{-}42)$$

$$k(-\sigma+j\omega+\sigma+j\omega)=\frac{V_0}{L} \qquad (8\text{-}43)$$

$$k=\frac{V_0}{2j\omega L} \qquad (8\text{-}44)$$

The time domain expression for the current equation may then be written as

8.4 OVERDAMPED CASE
The overdamped case is the case in which the actual circuit resistance is greater than the critical resistance. The damping ratio for this case is expressed by the following inequality:

$$\zeta > 1 \qquad (8\text{-}45)$$

The general expression for the roots of the characteristic equation are restated for convenience as follows:

TRANSIENT ANALYSIS

$$i(t) = \frac{V_0}{2j\omega L}\left[\varepsilon^{-\sigma+j\omega t} - \varepsilon^{-\sigma-j\omega t}\right]$$

$$= \frac{V_0}{2j\omega L}\left[\varepsilon^{-\sigma t}\varepsilon^{j\omega t} - \varepsilon^{-\sigma t}\varepsilon^{-j\omega t}\right]$$

$$= \frac{V_0}{\omega L}\varepsilon^{-\sigma t}\left[\frac{1}{2j}\varepsilon^{j\omega t} - \varepsilon^{-j\omega t}\right]$$

$$= \frac{V_0}{\omega L}\varepsilon^{-\sigma t}\sin\omega t$$

(8-46)

$$s_1, s_2 = -\omega_0\zeta \pm j\omega_0\sqrt{1-\zeta^2} \qquad (8\text{-}47)$$

The S-plane representation of the characteristic roots for the overdamped case is illustrated in Fig. 8-10. Note that the total current consists only of a magnitude component and that the magnitude decays exponentially due to the presence of losses.

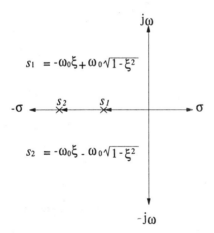

Figure 8-10 Characteristic roots for the overdamped case.

The time domain representation for the overdamped case is illustrated in Fig. 8-11. Note that the value of the resistor associated with the overdamped case is greater than the minimum value necessary to eliminate the oscillatory component.

CHAPTER 8

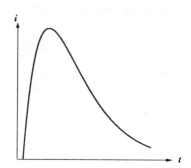

Figure 8-11 Time domain representation for the overdamped case.

Since the damping ratio is greater than unity, the characteristic roots s_1, s_2 will be real. The general expression for the roots of the characteristic equation may be restated as follows:

$$s_1, s_2 = -\omega_0\zeta \pm j\omega_0\sqrt{1-\zeta^2}$$
$$= -\omega_0\zeta \pm \omega_0\sqrt{\zeta^2-1} \qquad (8\text{-}48)$$

The characteristic roots for the overdamped case are summarized in Table 8-4.

Table 8-4 Characteristic Roots for the Overdamped Case

ROOTS	σ	$j\omega$
s_1	$-\omega_0\zeta + \omega_0\sqrt{\zeta^2-1}$	0
s_2	$-\omega_0\zeta - \omega_0\sqrt{\zeta^2-1}$	0

8.5 ENERGY CONSIDERATIONS

The initial conditions which were established for the RLC circuit were based upon an initial voltage V_0 across the capacitor. The energy stored in the electric field that exists in the dielectric of the capacitor is

$$W_C = \frac{1}{2}Cv^2 \qquad (8\text{-}49)$$

The closure of the switch at time permits the flow of charge from the plates of the capacitor through the inductor and the resistor. This current flow reaches a maximum value when all of the energy that was stored in the electric field of the capacitor has been transferred to the magnetic field which results from the flow of current through the inductance of the circuit. The voltage across the capacitor is reduced to zero at this point. The energy which is stored in the magnetic field is

$$W_L = \frac{1}{2}Li^2 \qquad (8\text{-}50)$$

The total energy in the circuit remains constant in the undamped or lossless case since there is no resistance to dissipated energy.

$$\frac{1}{2}Li^2 = \frac{1}{2}Cv^2 \qquad (8\text{-}51)$$

The addition of the resistor for the underdamped, critically damped, and overdamped cases results in the dissipation of energy in accordance with the relationships discussed in the previous sections.

$$\begin{aligned}W_{loss} &= \int p\,dt \\ &= \int Ri^2 dt\end{aligned} \qquad (8\text{-}52)$$

We know from Chap. 3 that a steady-state cosinusoidal current yields:

$$W_{loss} = RI^2 t \qquad (8\text{-}53)$$

where I is the rms value of current. This is an important consideration in power systems analysis. The heat energy that is dissipated by the flow of current through the resistance of a conductor may be substantial during fault conditions in which the short-circuit current can exceed rated current by 20 times. Manufacturers therefore

CHAPTER 8

impose limits on the I^2t values to which equipment may be exposed which are referred to as damage curves. Overcurrent devices such as fuses and overcurrent relays must be coordinated with the damage curves of protected equipment in order to ensure that equipment is disconnected prior to the onset of thermal damage. A typical damage curve for a transformer may be based on the following thermal limit:

$$I^2t = 1250 \tag{8-54}$$

The amount of time which a fault can persist at 25 times rated current before the equipment must be disconnected from service is

$$\begin{aligned} t &= \frac{1250}{I^2} \\ &= \frac{1250}{25^2} \\ &= 2 \text{ s} \end{aligned} \tag{8-55}$$

8.6 A PRACTICAL EXAMPLE

The telegrapher's equations developed in Chap. 4 yield the voltage and current on a transmission line at a given distance on the line and at a given point in time. We will use an approximate method to evaluate the transient response of a 765-kV transmission line with a 200-MVAR shunt reactor bank installed at both terminals to illustrate the application of the concepts developed in this chapter. The approximate T model for the circuit is illustrated in Fig. 8-12.

We will assume initial conditions such that the line capacitance upon simultaneous clearing of the power circuit breakers at the line terminals is charged to the full operating voltage of the system. The line-to-ground voltage across the line capacitance at the time of the trip is

$$V = \frac{765}{\sqrt{3}} = 442 \text{ kV} \tag{8-56}$$

The series impedance of the line is

$$\begin{aligned} Z &= R + jX_L \\ &= 2 + j70 \end{aligned} \tag{8-57}$$

TRANSIENT ANALYSIS

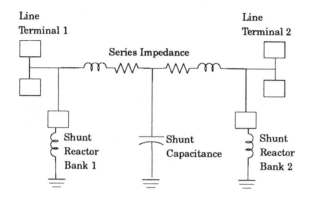

Figure 8-12 Transmission line model with shunt reactor banks.

The line charging at nominal voltage is 617 MVAR. The capacitive reactance of the line may be computed from the line charging and the nominal voltage as follows:

$$X_C = \frac{V^2}{Q}$$
$$= \frac{765^2}{617} \quad (8\text{-}58)$$
$$= 948.5 \; \Omega$$

The inductive reactance of the 200-MVAR shunt reactor banks may be computed as follows:

$$X_L = \frac{V^2}{Q}$$
$$= \frac{765^2}{200} \quad (8\text{-}59)$$
$$= 2926 \; \Omega$$

The inductive reactance of the shunt reactor banks and the transmission line may be combined as follows:

CHAPTER 8

$$X_L = j2926 + j35$$
$$= j2961 \ \Omega \quad (8\text{-}60)$$

The resolution of equal impedances in parallel yields

$$Z|| = \frac{Z}{n}$$
$$= \frac{1 + j2961}{2} \quad (8\text{-}61)$$
$$= 0.5 + j1480.5 \ \Omega$$

The inductance may be computed from the 60-Hz inductive reactance as follows:

$$L = \frac{\omega L}{\omega}$$
$$= \frac{1480.5}{377} \quad (8\text{-}62)$$
$$= 3.9 \ H$$

The capacitance may be computed from the 60-Hz capacitive reactance as follows:

$$C = \frac{1}{\omega X_C}$$
$$= \frac{1}{(377)(948.5)} \quad (8\text{-}63)$$
$$= 2.8 \ \mu F$$

The transmission line diagram of Fig. 8-12 may therefore be approximated in the same manner as the simple RLC circuit that was introduced in Fig. 8-1.

TRANSIENT ANALYSIS 179

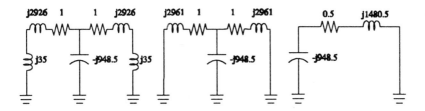

Figure 8-13 Successive reductions of the transmission line model.

We may now apply the theory that we have developed in the previous sections to evaluate the transient response of the network following the simultaneous clearing of the terminal breakers.

The undamped natural frequency is

$$\omega_0 = \frac{1}{\sqrt{LC}}$$
$$= \frac{1}{\sqrt{(3.9)(2.8)(10^{-6})}} \quad (8\text{-}64)$$
$$= 303 \text{ rad/s}$$

The critical resistance R_C is

$$R_C = 2\sqrt{\frac{L}{C}}$$
$$= 2\sqrt{\frac{3.9}{(2.8)(10^{-6})}} \quad (8\text{-}65)$$
$$= 2360 \ \Omega$$

The damping ratio is

$$\zeta = \frac{R}{R_C}$$
$$= \frac{0.5}{2360} \quad (8\text{-}66)$$
$$= 2.1 \times 10^{-4}$$

CHAPTER 8

The roots of the characteristic equation are

$$s_1, s_2 = -\omega_0\zeta \pm j\omega_0\sqrt{1-\zeta^2} \qquad (8\text{-}67)$$

Substituting the appropriate values yields

$$s_1, s_2 = -0.1 \pm j303 \qquad (8\text{-}68)$$

The time domain expression may be obtained from (8-46) in accordance with the particular solution for the underdamped case.

$$\begin{aligned} i(t) &= \frac{V_0}{\omega L}\varepsilon^{\sigma t}\sin\omega t \\ &= \frac{442}{(303)(3.9)}\varepsilon^{-0.1t}\sin 303t \\ &= 0.37\varepsilon^{-0.1t}\sin 303t \text{ kA} \end{aligned} \qquad (8\text{-}69)$$

Remember that the argument of the sinusoidal function is expressed in radians. Note that the frequency of oscillation of 303 rad/s is less than the scheduled power system frequency of 377 rad/s. The circuit time constant which represents the time in seconds for the current magnitude to decrease to 36.8 percent of its initial value is

$$\tau = \frac{L}{R} = \frac{3.9}{0.5} = 7.8 \text{ s} \qquad (8\text{-}70)$$

The reader will note that the response of the network is in accordance with the time domain representation of Fig. 8-9 for the underdamped case.

CHAPTER 9

SYMMETRICAL VERSUS ASYMMETRICAL CURRENT

Faults on electric power systems often involve short-circuit currents which are not symmetrical about the zero axis. The total current under these conditions is the sum of the transient and steady-state components.

9.1 INITIAL CONDITIONS

It is a characteristic of inductive circuits that the flux linkages and therefore the current cannot change instantaneously, e.g., in zero time. The voltage induced in an inductive circuit is based on the time rate of change of flux linkages.

Table 9-1 Components of Fault Current

Steady state	i_{ss}	This is the component of fault current that results from the application of the system voltage to the fault impedance.
Transient	i_T	This is the component of fault current that results from the initial conditions of voltage and current which existed at the time the fault occurred.
Total	$i = i_{ss} + i_T$	This is the total current which is the sum of the steady-state and transient components.

CHAPTER 9

$$v = d\frac{Li}{dt} \quad (9\text{-}1)$$

where L is the number of flux linkages per ampere and i is the number of amperes. The assumption of a constant inductance yields

$$v = L\frac{di}{dt} \quad (9\text{-}2)$$

Note that the change in current in zero time would result in an infinite voltage. This condition is often correlated with the relationship between force and the rate of change of linear momentum in mechanical systems.

$$F = d\frac{mv}{dt} \quad (9\text{-}3)$$

where m is the mass of an object, v is the velocity, t is the time, and dv/dt is the rate of change of velocity with respect to time which is also expressed as the acceleration a. The assumption of a constant mass yields

$$F = m\frac{dv}{dt} \quad (9\text{-}4)$$

Since $dv/dt = a$, then

$$F = ma \quad (9\text{-}5)$$

The mechanical force F is related to the rate of change of velocity with respect to time by the constant of proportionality m which represents the mass; similarly, the electromotive force v is related to the rate of change of current with respect to time by the constant of proportionality L which represents the inductance.

9.2 DERIVATION OF THE TOTAL CURRENT EQUATION

Consider the simplified model of a power system shown in Fig. 9-1. This system consists of generation, a system impedance, a transmission impedance, and a load impedance.

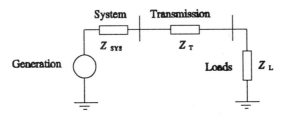

Figure 9-1 Simplified system model.

The circuit model for this system is illustrated in Fig. 9-2 and consists of a voltage source of the form $v=V_m\sin(\omega t+\theta_v)$, a phase current of the form $i=I_m\sin(\omega t+\theta_I)$, a series resistance R, and a series inductance L. The shunt parameters are neglected for this analysis.

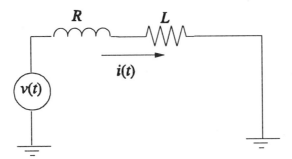

Figure 9-2 Circuit representation.

We will assume for purposes of this development that the bus under consideration is the reference bus so that the voltage angle is zero. The sum of the voltages around the loop therefore yields the following expression:

$$L\frac{di}{dt}+Ri=V_m\sin\omega t \qquad (9\text{-}6)$$

The equation may be alternately expressed as follows:

$$\frac{di}{dt}+\frac{R}{L}i=\frac{V_m}{L}\sin\omega t \qquad (9\text{-}7)$$

This is a first-order, linear, nonhomogeneous differential equation with constant coefficients. The solution for this equation consists of two

CHAPTER 9

components. The complementary solution i_C yields the transient component of current and is determined with the right side of the differential equation set equal to zero. The particular solution i_P yields the steady-state component and is determined with the right side of the equation equal to $V\sin\omega t$.

$$i = i_C + i_P \qquad (9\text{-}8)$$

9.2.1 Complementary Solution (i_c)

The complementary solution may be obtained by setting the right side of the equation equal to zero.

$$\frac{di}{dt} + \frac{R}{L}i = 0 \qquad (9\text{-}9)$$

$$\frac{di}{dt} = -\frac{R}{L}i \qquad (9\text{-}10)$$

$$\frac{di}{i} = -\frac{R}{L}dt \qquad (9\text{-}11)$$

$$\int \frac{di}{i} = -\int \frac{R}{L}dt \qquad (9\text{-}12)$$

$$\ln i = -\frac{R}{L}t + C \qquad (9\text{-}13)$$

$$i = \varepsilon^{-(R/L)t+C} = \varepsilon^{-(R/L)t} \varepsilon^C \qquad (9\text{-}14)$$

The time constant of the circuit which is measured in units of seconds is computed as follows:

$$\frac{L}{R} = \tau \qquad (9\text{-}15)$$

The initial current may be stated as follows:

$$I_0 = \varepsilon^C \qquad (9\text{-}16)$$

The complementary solution may therefore be stated as

$$i_C = I_0 \varepsilon^{-t/\tau} \quad (9\text{-}17)$$

9.2.2 Particular Solution (i_P)

The particular solution is chosen so as to include all possible values associated with the right side of the equation, e.g., the function and the first derivative of the function, as follows:

$$i_P = A\cos\omega t + B\sin\omega t \quad (9\text{-}18)$$

The first derivative of the particular solution is

$$\frac{di_P}{dt} = -\omega A \sin\omega t + \omega B \cos\omega t \quad (9\text{-}19)$$

Substituting di_P/dt and i_P into the original equation yields

$$(-\omega A\sin\omega t + \omega B\cos\omega t) + \frac{R}{L}(A\cos\omega t + B\sin\omega t) = \frac{V_m}{L}\sin\omega t \quad (9\text{-}20)$$

This process is the method of undetermined coefficients. Rearranging terms to equate coefficients on the left and right side of the equation yields

$$\sin\omega t\left(-\omega A + \frac{R}{L}B\right) + \cos\omega t\left(\frac{R}{L}A + \omega B\right) = \frac{V_m}{L}\sin\omega t \quad (9\text{-}21)$$

Equating terms yields the following simultaneous equations:

$$\begin{aligned} -\omega A + \frac{R}{L}B &= \frac{V_m}{L} \\ \frac{R}{L}A + \omega B &= 0 \end{aligned} \quad (9\text{-}22)$$

Solving for A and B yields

$$\begin{aligned} A &= -\frac{\omega L V_m}{R^2 + (\omega L)^2} \\ B &= \frac{R V_m}{R^2 + (\omega L)^2} \end{aligned} \quad (9\text{-}23)$$

The particular solution may now be expressed as follows:

CHAPTER 9

$$i_P = A\cos\omega t + B\sin\omega t$$

$$= -\left[\frac{\omega L V_m}{R^2+(\omega L)^2}\right]\cos\omega t + \left[\frac{RV_m}{R^2+(\omega L)^2}\right]\sin\omega t \qquad (9\text{-}24)$$

$$= \frac{V_m}{R^2+(\omega L)^2}(R\sin\omega t - \omega L\cos\omega t)$$

Note that by inspection the following expressions can be established:

$$\begin{aligned}Z^2 &= R^2+(\omega L)^2 \\ Z &= \sqrt{R^2+(\omega L)^2} \\ R &= Z\cos\theta \\ \omega L &= Z\sin\theta\end{aligned} \qquad (9\text{-}25)$$

The particular solution may be expressed as follows:

$$\begin{aligned}i_P &= \frac{V_m}{Z^2}(Z\cos\theta\sin\omega t - Z\sin\theta\cos\omega t) \\ &= \frac{V_m}{Z}(\cos\theta\sin\omega t - \sin\theta\cos\omega t)\end{aligned} \qquad (9\text{-}26)$$

This expression may be further reduced by the following trigonometric identity:

$$\sin\alpha\cos\beta - \cos\alpha\sin\beta = \sin(\alpha-\beta) \qquad (9\text{-}27)$$

where $\alpha=\omega t$ and $\beta=\theta$. The particular solution may therefore be expressed as follows:

$$i_P = \frac{V_m}{\sqrt{R^2+(\omega L)^2}}\sin(\omega t-\theta) \qquad (9\text{-}28)$$

9.2.3 Complete Solution ($i=i_P+i_C$)

The complete solution is the sum of the complementary solution and the particular solution.

$$\begin{aligned}i &= i_P+i_C \\ &= \frac{V_m}{\sqrt{R^2+(\omega L)^2}}\sin(\omega t-\theta) + I_0\varepsilon^{-(R/L)t}\end{aligned} \qquad (9\text{-}29)$$

The current at $t=0$ is

$$i = \frac{V_m}{\sqrt{R^2+(\omega L)^2}}\sin(-\theta)+I_0 \qquad (9\text{-}30)$$

We will consider initial conditions in which the instantaneous value of prefault current which was flowing on the circuit just prior to the fault is equal to zero, e.g., $i(0-)=0$. Since the current flow cannot change instantaneously in an inductive circuit, then $i(0-)=i(0+)$. The sum of the steady-state and transient components of fault current must then equal the prefault current which was flowing on the circuit just prior to the event which in this case is zero. The equation may then be rewritten as follows:

$$i(0)=i_{SS}+i_T$$
$$= \frac{V_m}{\sqrt{R^2+(\omega L)^2}}\sin(-\theta)+I_0 \qquad (9\text{-}31)$$
$$=0$$

$$I_0 = -\frac{V_m}{\sqrt{R^2+(\omega L)^2}}\sin(-\theta) \qquad (9\text{-}32)$$

Absorbing the minus sign into the argument yields

$$I_0 = \frac{V_m}{\sqrt{R^2+(\omega L)^2}}\sin(\theta) \qquad (9\text{-}33)$$

The complete solution then becomes

$$i = \frac{V_m}{\sqrt{R^2+(\omega L)^2}}\sin(\omega t-\theta)+\varepsilon^{-t/\tau}\left[\frac{V_m}{\sqrt{R^2+(\omega L)^2}}\sin(\theta)\right] \qquad (9\text{-}34)$$

9.3 FAULT ANALYSIS

We will now conduct a fault analysis on the 345-kV system shown in Fig. 9-3 to determine the magnitude of the fault current out to five cycles.

CHAPTER 9

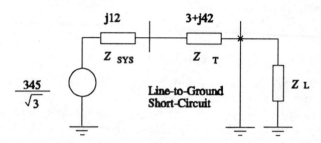

Figure 9-3 Faulted system.

The line-to-ground voltage applied at the fault point is

$$V = \frac{345}{\sqrt{3}} = 199 \text{ kV} \quad (9\text{-}35)$$

The system impedance is

$$Z = j12 \text{ } \Omega \quad (9\text{-}36)$$

The impedance of the transmission line is

$$Z = R + j\omega L = 3 + j42 \text{ } \Omega \quad (9\text{-}37)$$

The total fault impedance is:

$$Z = R + j\omega L = 3 + j(42+12) = 54.1 \angle 86.8° \quad (9\text{-}38)$$

The inductance of the system is

$$L = \frac{\omega L}{\omega} = \frac{54}{377} = 0.14 \text{ H} \quad (9\text{-}39)$$

The time constant of the circuit is

$$\tau = \frac{L}{R} = \frac{0.14}{3} = 0.05 \text{ s} \quad (9\text{-}40)$$

We restate the expression for the total fault current as the sum of the steady-state and transient components as follows:

$$i = \frac{V_m}{\sqrt{R^2 + (\omega L)^2}} \sin(\omega t - \theta) + \varepsilon^{-t/\tau} \left[\frac{V_m}{\sqrt{R^2 + (\omega L)^2}} \sin(\theta) \right] \quad (9\text{-}41)$$

SYMMETRICAL VERSUS ASYMMETRICAL CURRENT

The total fault current and the steady-state and transient components are illustrated in Figure 9-4.

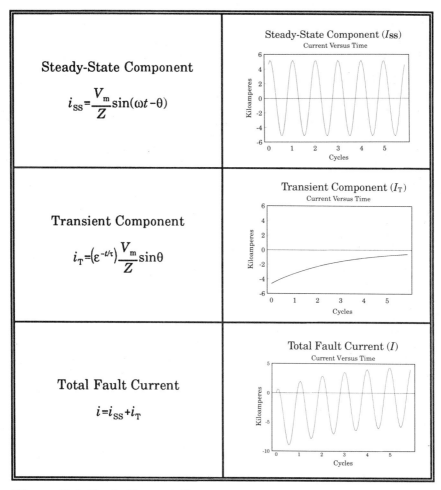

Figure 9-4 Fault Current Components.

CHAPTER 10

TRANSFORMERS

This discussion focuses on a variety of transformers that are used in the electric utility industry including power and instrument transformers. It is helpful to review the fundamental theory of transformers as an aid to our investigation of transformer applications.

10.1 BASIC TRANSFORMER THEORY
The discussion which follows is based on two-winding transformers, e.g., transformers which have a primary and a secondary winding.

The primary voltage and current and the number of turns in the primary winding are e_1, i_1, and n_1, respectively; similarly, the secondary voltage and current and the number of turns of the secondary winding are e_2, i_2, and n_2.

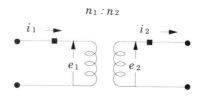

Figure 10-1 Basic transformer model.

The polarity marks which are the square boxes on the primary and secondary windings indicate that the current i_1 which enters the polarity mark on the primary winding is essentially in phase with the current i_2 which flows out of the polarity mark on the secondary winding.[12]

---

[12]The term "essentially" is used because the R/L nature of a transformer causes a phase shift which is measured in minutes or 1/60 of a degree between the primary and secondary circuits. A phase angle correction factor is provided to account for this discrepancy.

10.2 TRANSFORMER FLUX RELATIONSHIPS

The magnetic fluxes that result from the flow of current through the windings of a transformer are categorized in terms of mutual flux Φ_m and leakage flux Φ_l. Mutual flux is that portion of the total flux which links both the primary and secondary windings. Leakage flux is that portion of the total flux generated by each winding which does not link the other winding. This concept can be illustrated most effectively by considering the fluxes produced by the primary and secondary currents separately.

The primary leakage flux Φ_{11} is the component of the primary flux that is generated by the primary current and which does not link the secondary winding. The self-inductance L_{11} of the primary winding represents the number of lines of primary leakage flux that link the primary winding per ampere of primary current.

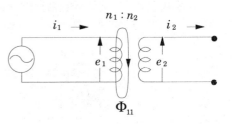

Figure 10-2 Primary leakage flux.

$$L_{11} = \frac{\Phi_{11}}{i_1}$$

The secondary leakage flux Φ_{22} is the component of the secondary flux that is generated by the secondary current and which does not link the primary winding. The self-inductance of the secondary winding represents the number of lines of secondary leakage flux that link the secondary winding per ampere of secondary current.

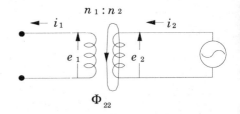

Figure 10-3 Secondary leakage flux.

$$L_{22} = \frac{\Phi_{22}}{i_2}$$

The mutual flux Φ_{21} is the component of the primary flux that is generated by the primary current and which also links the secondary winding. The mutual inductance L_{21} between the primary and secondary windings represents the number of lines of primary flux that link the secondary winding per ampere of primary current.

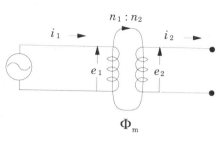

Figure 10-4 Primary mutual flux.

$$L_{21} = \frac{\Phi_{21}}{i_1}$$

The mutual flux Φ_{12} is the component of the secondary flux that is generated by the secondary current and which links both the primary and secondary windings. The mutual inductance L_{12} between the primary and secondary windings represents the number of lines of secondary flux that link the primary winding per ampere of secondary current.

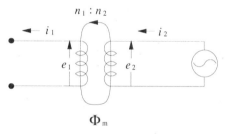

Figure 10-5 Secondary mutual flux.

$$L_{12} = \frac{\Phi_{12}}{i_2}$$

It can therefore be established that the primary and secondary fluxes consist of mutual and leakage fluxes as summarized in Table 10-1.

CHAPTER 10

Table 10-1 Summary of Mutual and Leakage Flux

WINDING FLUX	DUE TO PRIMARY CURRENT	DUE TO SECONDARY CURRENT
Primary flux Φ_1	Φ_{11}	Φ_{12}
Secondary flux Φ_2	Φ_{21}	Φ_{22}

10.3 THE IDEAL TRANSFORMER

The ideal transformer which is illustrated in Fig. 10-6 has no resistance and therefore no real power losses. In addition, all of the flux in the transformer is mutual flux; e.g., there is no leakage flux associated with either winding.

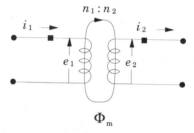

Figure 10-6 The ideal transformer.

The mutual flux between the primary and secondary windings is Φ_m. The primary flux linkages Ψ_1 are based on the number of lines of flux Φ_m that link the primary winding and the number of turns n_1 of the primary winding.

$$\Psi_1 = n_1 \Phi_m \qquad (10\text{-}1)$$

The secondary flux linkages Ψ_2 are based on the number of lines of flux Φ_m that link the secondary winding and the number of turns n_2 of the secondary winding. The ideal transformer is an important device since it allows us to simplify the model of a real transformer.

$$\Psi_2 = n_2 \Phi_m \qquad (10\text{-}2)$$

10.4 TRANSFORMER ACTION

Consider the ideal transformer shown in Fig. 10-7. A voltage source is applied to the primary winding, and the secondary winding is open-circuited. The current in the primary winding produces lines of flux which link the secondary winding. A sinusoidal variation in the primary current i_1 will produce a sinusoidal variation in the mutual flux linkages Ψ_m. This time-varying flux will induce a voltage e_2 in the secondary winding of the ideal transformer which is proportional to the time rate of change of flux linkages.

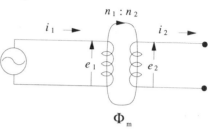

Figure 10-7 Induced voltage.

$$e_2 = \frac{d(n_2\Phi_m)}{dt} = \frac{d\Psi_2}{dt} \quad (10\text{-}3)$$

The flux linkages under sinusoidal conditions change four times between zero and maximum ($n\Phi_{max}$) during the period of each cycle:

- From zero to maximum in the positive direction in the first quarter cycle.

- From maximum in the positive direction to zero in the second quarter cycle.

- From zero to maximum in the negative direction in the third quarter cycle

- From maximum in the negative direction to zero in the fourth quarter cycle.

The total change in flux linkages over the interval of a complete cycle is $4n\Phi_{max}$. The average voltage is then equal to the average change in flux linkages per period or four times the change in flux linkages in each quarter period divided by the total period per cycle.

$$E_{ave}=4n\Phi_{max}\left(\frac{1}{\tau}\right) \qquad (10\text{-}4)$$

The reciprocal of the period is the applied frequency f, and the equation may be expressed as follows:

$$E_{ave}=4n\Phi_{max}f \qquad (10\text{-}5)$$

The effective or rms value may be computed by division of the average value by 0.637 to obtain the maximum value and by multiplication of the maximum value by 0.707 to obtain the rms value. This results in the following form factor:

$$\frac{0.707}{0.637}=1.11 \qquad (10\text{-}6)$$

The rms voltage is then computed as follows:

$$E_{rms}=\left(\frac{0.707}{0.637}\right)4n\Phi_{max}f \qquad (10\text{-}7)$$

The rms value of quantities is assumed in this material unless otherwise stated. The rms voltage may therefore be expressed as

$$E=4.44n\Phi_{max}\,f \qquad (10\text{-}8)$$

The maximum flux may be stated as

$$\Phi_{max}=\frac{E}{4.44nf} \qquad (10\text{-}9)$$

The maximum flux linkages may be obtained by multiplying both sides of the equation by the number of turns.

$$\Psi_{max}=\frac{E}{4.44f} \qquad (10\text{-}10)$$

The flux density B in webers per unit area may be obtained by dividing both sides of the equation by the cross-sectional area of the flux path A.

TRANSFORMERS

$$B_{max} = \frac{E}{4.44 nAf} \qquad (10\text{-}11)$$

The maximum flux density may be expressed in terms of lines per square inch by multiplying the webers per square inch by 10^8.

10.5 TURNS RATIO
The voltages induced in the primary and secondary windings are computed as follows:

$$e_1 = n_1 \frac{d\Psi_m}{dt} \qquad (10\text{-}12)$$

$$e_2 = n_2 \frac{d\Psi_m}{dt} \qquad (10\text{-}13)$$

The ratio of the primary to the secondary voltage yields the following:

$$\frac{e_1}{e_2} = \frac{n_1 \left(\dfrac{d\Phi_m}{dt}\right)}{n_2 \left(\dfrac{d\Phi_m}{dt}\right)} = \frac{n_1}{n_2} \qquad (10\text{-}14)$$

The turns ratio a which is the ratio of the number of primary turns to the number of secondary turns may be related to the primary and secondary voltages as follows:

$$\frac{n_1}{n_2} = \frac{e_1}{e_2} = a \qquad (10\text{-}15)$$

10.6 COEFFICIENT OF COUPLING
The coefficient of coupling k is a measure of the relationship between the mutual inductance M between the primary and secondary windings and the self-inductances L_{11} and L_{22}.

The ratio for the primary winding is

$$\frac{\Phi_{12}}{\Phi_1} \qquad (10\text{-}16)$$

The ratio for the secondary winding is

$$\frac{\Phi_{21}}{\Phi_2} \qquad (10\text{-}17)$$

The coefficient of coupling is computed as the geometric mean of the ratios for each winding as follows:

$$k = \sqrt{\left(\frac{\Phi_{12}}{\Phi_1}\right)\left(\frac{\Phi_{21}}{\Phi_2}\right)} \qquad (10\text{-}18)$$

Since $\Phi_{12}=\Phi_{21}=M$, then the coefficient of coupling may be expressed as

$$k = \sqrt{\frac{\Phi_M^2}{\Phi_1 \Phi_2}} \qquad (10\text{-}19)$$

The turns ratio a is computed on the basis that all of the flux in the transformer is mutual flux and that the coefficient of coupling is equal to unity. The presence of leakage flux yields a ratio of primary to secondary voltage that is not in accordance with the turns ratio. A ratio correction factor (RCF) is therefore provided to represent the true voltage ratio of the transformer. In the interest of efficiency, the modern power transformer used in the electric utility industry has a coefficient of coupling that approaches unity. In addition, the modern instrument transformer also has a high coefficient of coupling in order to achieve the desired level of accuracy for the measurement of voltage and current.

10.7 PRIMARY AND SECONDARY QUANTITIES
The usefulness of the ideal transformer in the development of circuit models for real transformers will become evident in the following development. The ideal transformer is assumed to be lossless. The apparent power is therefore the same in the primary and secondary windings.

TRANSFORMERS

$$S_1 = S_2 \tag{10-20}$$

This relationship may be restated in order to obtain the relationship between the primary and secondary current.

$$S_1 = S_2$$
$$E_1 I_1 = E_2 I_2 \tag{10-21}$$
$$\frac{E_1}{E_2} = \frac{I_2}{I_1} = a$$

An inverse relationship therefore exists between the ratio of primary voltage to secondary voltage on the one hand and primary current to secondary current on the other hand. The impedance relationships may also be developed in this manner.

$$S_1 = S_2$$
$$\frac{E_1^2}{Z_1} = \frac{E_2^2}{Z_2} \tag{10-22}$$
$$\frac{E_1^2}{E_2^2} = \left(\frac{E_1}{E_2}\right)^2 = a^2 = \frac{Z_1}{Z_2}$$

This gives rise to the following important relationships:

$$Z_1 = a^2 Z_2 \tag{10-23}$$

$$Z_2 = \frac{Z_1}{a^2} \tag{10-24}$$

We may now examine the adjustments that must be made to the ideal transformer in order to represent a practical transformer.

10.8 THE PRACTICAL TRANSFORMER
The circuit model of a practical transformer is shown in Fig. 10-8.

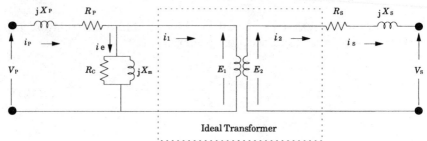

Figure 10-8 Practical transformer model.

10.9 TRANSFORMER IMPEDANCES
The series impedances of the primary and secondary windings are stated as follows:

$$Z_P = R_P + jX_{lP} \qquad (10\text{-}25)$$

$$Z_S = R_S + jX_{lS} \qquad (10\text{-}26)$$

10.9.1 Winding Resistance
The resistances R_P and R_S shown in the primary and secondary circuits of the diagram represent the winding resistances. The power transformer has real power or I^2R losses associated with the flow of current through the transformer windings. These losses are sometimes referred to as copper losses.

10.9.2 Winding Reactance
The reactances X_P and X_S shown in the primary and secondary circuits of the diagram represent the leakage reactances in the windings. The power transformer has reactive power or I^2X losses associated with the leakage flux between the windings.

10.9.3 Core Resistance
The core resistance R_C represents the resistance to eddy currents which flow in the core material and to the currents that flow as the result of hystereses. This circuit element contributes to the real power loss in the transformer. The core of the power transformer is

TRANSFORMERS

constructed from laminated iron which serves to reduce the flow of eddy currents and hence the core losses. The core resistance is proportional to the resistivity of the core material and the length of the core path and inversely proportional to the cross-sectional area of the magnetic circuit.

$$R_C \propto \rho_C \frac{l}{A} \qquad (10\text{-}27)$$

10.9.4 Magnetizing Reactance

The application of a sinusoidal voltage to a winding of the transformer causes a magnetizing or exciting current to flow which induces a flux in the core. The magnetizing inductance represents the number of lines of flux created in the core per ampere of magnetizing current. The magnetizing reactance X_m is determined as follows:

$$X_m = \omega L_m \qquad (10\text{-}28)$$

The application of the system voltage to the winding terminals of a transformer may result in a magnetizing current which is between 2 and 4 percent of the rated current of the transformer.

10.10 REFLECTED PRIMARY AND SECONDARY QUANTITIES

The transformer model can be reduced by reflecting all quantities to either the primary or secondary side of the transformer.

10.10.1 Secondary Quantities Reflected to the Primary

The secondary quantities may be reflected into the primary side of the transformer. The primary quantities are restated as follows:

$$Z_P = R_P + jX_{lP} \qquad (10\text{-}29)$$

The secondary quantities reflected into the primary yield

$$\begin{aligned} Z_P' &= a^2 Z_S \\ &= a^2 (R_S + jX_S) \\ &= a^2 R_S + ja^2 X_S \end{aligned} \qquad (10\text{-}30)$$

The circuit models for this condition are illustrated in Figs. 10-9 and 10-10.

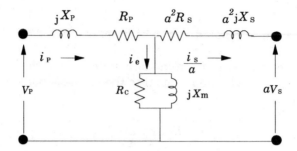

Figure 10-9 Transformer circuit model with secondary impedance reflected into the primary (R/X model).

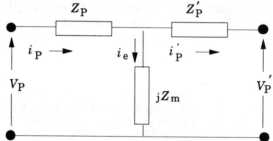

Figure 10-10 Transformer circuit model with secondary impedance reflected into the primary (Z model).

10.10.2 Primary Quantities Reflected to the Secondary

The primary quantities may also be reflected into the secondary side of the transformer. The secondary quantities are restated as follows:

$$Z_S = R_S + jX_{lS} \qquad (10\text{-}31)$$

The primary quantities reflected into the secondary yield

$$\begin{aligned} Z_S' &= \frac{Z_P}{a^2} \\ &= \frac{R_P + jX_{lP}}{a^2} \\ &= \frac{R_P}{a^2} + j\frac{X_{lP}}{a^2} \end{aligned} \qquad (10\text{-}32)$$

The circuit models for this condition are illustrated in Figs. 10-11 and 10-12.

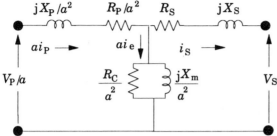

Figure 10-11 Transformer circuit model with primary impedance reflected into the secondary (R/X model).

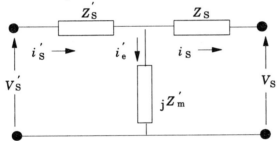

Figure 10-12 Transformer circuit model with primary impedance reflected into the secondary (Z model).

10.11 TRANSFORMER EXCITATION

The core and shell configurations are typically used in the design of transformers.

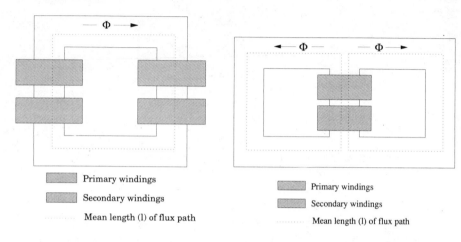

Figure 10-13 Core-type transformer.

Figure 10-14 Shell-type transformer.

Iron is typically used as a core material in a practical transformer. The relative permeability μ_r of free space is taken to be equal to unity. The relative permeability of iron is on the order of 10,000. The magnetic flux density B in the core of the transformer is a measure of the number of lines of flux per unit area in the core.

$$B = \frac{\Phi}{A} \tag{10-33}$$

The magnetic flux intensity may be expressed as the magnetizing current i_m per unit length of the flux path l.

$$H = \frac{i_m}{l} \tag{10-34}$$

TRANSFORMERS

The magnetic flux density as a function of the magnetic flux intensity H and the permeability μ of the core material may be represented as the hystereses or BH curve.

$$B = \mu H$$

The residual flux B_0 and the coercive intensity H_C are represented on the diagram.

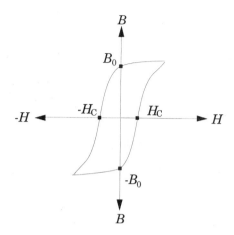

Figure 10-15 Hystereses (BH) curve.

Note that the permeability is the slope of the BH curve.

$$\mu = \frac{B}{H} \tag{10-35}$$

Let us restate that the flux density is the measure of the number of lines of flux per unit area.

$$B = \frac{\Phi}{A} \tag{10-36}$$

In addition, the magnetizing force is based on the magnetic flux intensity and the mean length of the flux path.

$$F = H \cdot l = \frac{ni}{l} \cdot l = ni \tag{10-37}$$

CHAPTER 10

The length and cross-sectional area of the core and the number of turns in the winding are constant. A relationship may therefore be developed between the number of lines of magnetizing flux Φ and the magnetizing current i_e. Note that the slope of this relationship is the inductance of the circuit.

$$L = \frac{\Phi}{i}$$

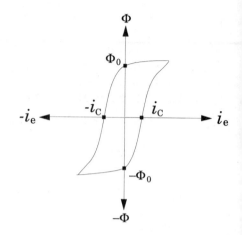

Figure 10-16 Flux versus magnetizing current.

This confirms our assertion that the inductance of the transformer circuit is a measure of the number of lines of flux per ampere of magnetizing current. We are now prepared to discuss the effects of saturation of the transformer core.

10.12 SATURATION
The flux density increases linearly in direct proportion to the magnetizing current until the point of magnetic saturation is reached. The maximum flux density for the iron core material is typically on the order of 100,000 lines per square inch. Further increases in magnetizing current beyond the point of magnetic saturation will force the lines of flux into the air surrounding the winding which has a significantly lower permeability. This results in a sharp decrease in the number of lines of flux per ampere of current which in turn represents a sharp decrease in the inductance of the winding. Since the magnetizing reactance is proportional to the inductance of the magnetizing circuit, then a sharp decrease in magnetizing inductance results in a sharp decrease in the magnetizing reactance. The magnetizing circuit then tends toward a short circuit under the condition of magnetic saturation.

10.13 DELTA-WYE TRANSFORMATIONS

Delta windings may be used in generator step-up transformers to connect generators to the electric power systems since they represent an open circuit in the zero sequence network for faults on the system side of the transformer. In addition, they block the flow of third-harmonic currents on the system. The following discussion illustrates the connection of a power transformer in which the primary winding terminals X_1, X_2, X_3 are connected in a delta configuration and the secondary winding terminals H_1, H_2, H_3 are connected in a grounded-wye configuration.

10.13.1 Current Relationships

The current relationships for the delta and wye connections are based on Kirchoff's current law which states that the currents which enter a node must sum to zero. The connections of the phase conductors to the wye and delta windings represent current nodes.

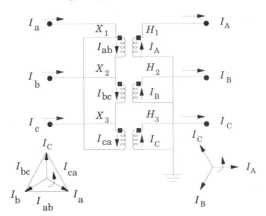

Figure 10-17 Delta-wye currents.

10.13.1.1 Delta Currents.
Each of the three current nodes X_1, X_2, X_3 on the delta side is connected to three branches. Two of the branches are the adjacent windings, and the other branch is the associated phase conductor. The application of Kirchoff's current law requires that the current which flows through the phase conductor on each phase of the delta connection is equal to the vector sum of the currents which flow through the adjacent windings. The relationships between

the phase currents I_a, I_b, I_c and the winding currents I_{ab}, I_{bc}, I_{ca} may be determined by inspection of the currents which flow into each of the three current nodes X_1, X_2, X_3. These relationships are summarized in Table 10-2.

Table 10-2 Summary of Delta Currents

DELTA CONNECTION	RELATIONSHIPS	PHASOR DIAGRAM
X_1	$I_a = I_{ab} - I_{ca}$	
X_2	$I_b = I_{bc} - I_{ab}$	
X_3	$I_c = I_{ca} - I_{bc}$	

The relationships between the winding currents and the phase currents on the delta side are shown in Table 10-3.

Table 10-3 Delta Phase and Winding Currents

DELTA PHASE CURRENTS	DELTA WINDING CURRENTS

10.13.1.2 Wye Currents. Each of the three current nodes H_1, H_2, H_3 on the wye side is connected to two branches. One branch is the phase conductor, and the other branch is the associated winding. The application of Kirchoff's current law requires that the current which flows through the phase conductor on each phase of the wye connection is equal to the current which flows through the associated winding. The resultant current relationships are illustrated in Table 10-4.

Table 10-4 Wye Phase and Winding Currents

WYE PHASE CURRENTS	WYE WINDING CURRENTS
I_C, I_A, I_B (phasor diagram)	I_C, I_A, I_B (phasor diagram)

The current node H_0 is connected to four branches. Three of the branches are the nonpolarity connections of the wye windings, and the fourth is the grounded neutral conductor. This permits the flow of residual currents to ground during unbalanced conditions.

10.13.1.3 Delta-Wye Phase Shift for Currents. The foregoing information is summarized by the phasor representations of the delta phase currents and the wye phase currents (Table 10-5). Note that the delta phase currents lag the corresponding wye phase currents by 30°.

Table 10-5 Delta and Wye Phase Currents

WYE PHASE CURRENTS	WYE WINDING CURRENTS
I_c, I_b, I_a (phasor diagram)	I_C, I_A, I_B (phasor diagram)

Alternate phase shifts may be obtained by various connections of the winding terminals. The reader is encouraged to review the reference material for this chapter and to determine the appropriate connections for other degrees of phase shift. The polarity markings which are used in the electric utility industry regarding the phase relationships for transformers are defined in the ANSI/IEEE Standard No. C57.12.70 entitled, "Terminal Markings and Connections for Distribution and entitled, "Terminal Markings and Connections for Distribution and Power Transformers." The standard convention states that the high-side voltage should lead the low-side voltage on each phase regardless of whether the transformation is for a delta-wye or wye-delta connection.

10.13.2 Voltage Relationships

The voltage relationships for the delta and wye connections are based on Kirchoff's voltage law which states that the voltages around a closed loop must sum to zero. The connections of the windings to the phase conductors on the wye and delta sides form voltage loops.

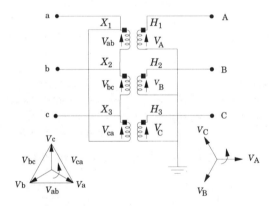

Figure 10-18 Delta-wye voltages.

TRANSFORMERS   211

10.13.2.1 Delta Voltages. The three winding voltages on the delta side are V_{ab}, V_{bc}, V_{ca}. These winding voltages form the phasor triangle in the delta-wye circuit diagram for voltages shown in Fig. 10-18. The delta phase and winding voltages are summarized in Table 10-6.

Table 10-6 Delta Phase and Winding Voltages

DELTA PHASE VOLTAGES	DELTA WINDING VOLTAGES
V_c V_b V_a	V_{ca} V_{ab} V_{bc}

10.13.2.2 Wye Voltages. The three phase voltages on the wye side are V_A, V_B, V_C. These voltages appear across each of the wye-connected windings and are summarized in Table 10-7.

Table 10-7 Wye Phase and Winding Voltages

WYE PHASE VOLTAGES	WYE WINDING VOLTAGES
V_C V_A V_B	V_C V_A V_B

10.13.2.3 Delta-Wye Phase Shift for Voltages. The foregoing information is summarized in Table 10-8 by the phasor representations of the delta phase voltages and the wye phase voltages. Note that the delta phase voltages lag the wye phase voltages by 30°.

Table 10-8 Wye Phase and Winding Voltages

DELTA PHASE VOLTAGES	WYE PHASE VOLTAGES
V_c, V_b, V_a	V_C, V_A, V_B

10.14 CURRENT TRANSFORMERS

Current transformers are used for the measurement of current on the phase and ground conductors of the electric power system. These currents may be applied for such applications as revenue metering, indication, and protective relaying. A functional diagram of an overcurrent protection scheme for a circuit is shown in Fig. 10-19. The primary winding of a current transformer has one turn which represents the conductor or conductors being measured. The primary winding must be rated for the normal, emergency, and fault current that it will be expected to carry. A typical current transformer ratio for the bulk electric power system is 2000:5 which yields 5 secondary amperes for every 2000 primary amperes.

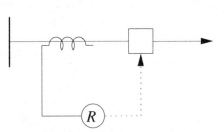

Figure 10-19 Current transformer, overcurrent relay, and power circuit breaker.

It is necessary to ensure that the secondary winding of the current transformer does not become open-circuited due to the inversion of the voltage and current ratios and the resultant high voltage that would be produced across the many turns of the open-circuited secondary winding. Shorting blocks are provided in the current circuits of protective relays in order to interrupt the current flow through them from the current transformer.

10.14.1 Current Transformer Model
A circuit model of a current transformer is shown in Fig. 10-20 with all impedances reflected into the secondary.

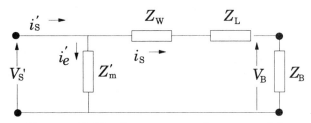

Figure 10-20 Current transformer model.

10.14.2 Burden
The burden of the current transformer is the impedance in ohms that is connected to the secondary terminals. The total burden is comprised of the current transformer impedance, the connecting leads, and the protective relays.

The relay burden is the impedance of the relay coils that are connected in series with the secondary winding and which receive the secondary current. The lead burden is the impedance of the connecting leads between the current transformer which may be located in an outdoor switchyard and the protective relays which may be located inside of a service and relay (S&R) building. The current transformer impedance is the impedance of the primary and secondary windings. We will assume a current transformer burden as summarized in Table 10-9.

Table 10-9 Current Transformer Burden

WINDING IMPEDANCE	LEAD IMPEDANCE	RELAY BURDEN	TOTAL CT BURDEN
$Z_W = 1\ \Omega$	$Z_L = 1\ \Omega$	$Z_R = 1\ \Omega$	$Z_T = 1\ \Omega$

10.14.3 Performance Under Normal Conditions
The secondary voltage that appears across the magnetizing branch of the current transformer is computed from the IZ drop across the secondary terminals. The secondary voltage at the full rated current

of 5 A is as follows:

$$V_S = I_S Z_S$$
$$= I_S(Z_W + Z_L + Z_R) \quad (10\text{-}38)$$
$$= 5(1+1+2)$$
$$= 20 \text{ V}$$

The exciting current through the magnetizing branch of the current transformer will be a linear function of the secondary voltage below the point of saturation since the inductance is linear in this region.

10.14.4 Performance Under Fault Conditions

Now consider that the maximum anticipated fault current is determined on the basis of short-circuit studies to be 40,000 A. This represents a fault current level of 20 times rated current. The transformer turns ratio of 2000:5 A yields a secondary current of 100 A. The secondary voltage is now computed as follows:

$$V_S = I_S Z_S$$
$$= I_S(Z_W + Z_L + Z_R) \quad (10\text{-}36)$$
$$= 100(1+1+2)$$
$$= 400 \text{ V}$$

It is necessary that the current transformer does not saturate during fault conditions in order that the secondary current to the protective relays will be an accurate reflection of the primary current. A secondary voltage that is beyond the rating of the current transformer will produce excessive levels of magnetizing current which will in turn cause a level of flux that will saturate the core.

The rapid decline in the inductance of the magnetizing branch causes a corresponding decline of the magnetizing reactance. The magnetizing branch therefore tends toward a short circuit and the secondary current is diverted from the protective relays. A current transformer accuracy class of C400 with a standard burden of 4 Ω is required for this application. The current transformer accuracy classes are illustrated in Fig. 10-21.

TRANSFORMERS

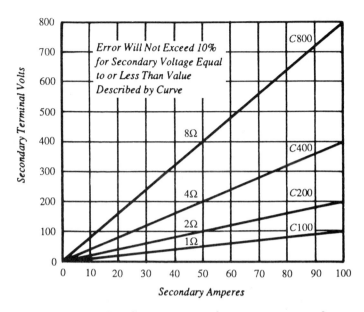

Figure 10-21 Current transformer accuracy class.

APPENDIX

SOLUTION METHODS FOR TRANSMISSION LINE EQUATIONS

A.1 METHOD OF LAPLACE TRANSFORM

A.1.1 Solution of the Voltage Equation

We will first consider the solution of the voltage equation which is restated for convenience as follows:

$$\frac{d^2V}{dx^2} = \gamma^2 V \qquad (A-1)$$

The Laplace transform of the voltage as a function of distance is expressed as follows:

$$\mathcal{L}V(x) = V(s) \qquad (A-2)$$

Taking the Laplace transform of the voltage equation yields

$$s^2 V(s) - sV(0+) - V'(0+) = \gamma^2 V(s) \qquad (A-3)$$

The terms $V(0+)$ and $V'(0+)$ are initial conditions. They represent the constant terms which were present prior to the two successive differentiations of the original equation to produce the second-order voltage equation. The first initial condition $V(0+)$ represents the voltage at the source terminal of the transmission line ($x=0$) which we will define as follows:

$$V(0+) = V_0 \qquad (A-4)$$

The second initial condition $V'(0+)$ represents the rate of change of voltage with respect to distance along the transmission line at the source terminal. We will assume no discontinuity of the transmission line at the source terminal so that $V(0+)$ becomes $V(0)$ and $V'(0+)$ becomes $V'(0)$.

APPENDIX

The rate of change of voltage with respect to distance along the transmission line was stated earlier to be

$$\frac{dV}{dx} = -IZ \qquad \text{(A-5)}$$

The current at the source terminal of the transmission line is defined as

$$I(0) = I_0 \qquad \text{(A-6)}$$

The rate of change of voltage at the source terminal of the transmission line is therefore

$$\frac{dV}{dx} = -I_0 Z \qquad \text{(A-7)}$$

The initial conditions are then summarized as follows:

$$V(0) = V_0 \qquad \text{(A-8)}$$

$$V'(0) = -I_0 Z \qquad \text{(A-9)}$$

The Laplace transform of the voltage equation with the substitution of initial conditions can now be expressed as follows:

$$s^2 V(s) - sV_0 + I_0 Z = \gamma^2 V(s) \qquad \text{(A-10)}$$

Manipulating the equation to solve for $V(s)$ yields

$$V(s)(s^2 - \gamma^2) = sV_0 - I_0 Z \qquad \text{(A-11)}$$

$$V(s) = \frac{sV_0 - I_0 Z}{s^2 - \gamma^2} \qquad \text{(A-12)}$$

$$V(s) = V_0 \left(\frac{s}{s^2 - \gamma^2} \right) - \frac{I_0 Z}{\gamma} \left(\frac{\gamma}{s^2 - \gamma^2} \right) \qquad \text{(A-13)}$$

SOLUTION METHODS

The solution of the voltage equation as a function of distance $V(x)$ may be found by the inverse Laplace transform of the function of s as follows:

$$\mathcal{L}^{-1}[V(s)] = V(x) = V_0 \cosh(\gamma x) - \frac{I_0 Z}{\gamma} \sinh(\gamma x) \quad \text{(A-14)}$$

A.1.2 Solution of the Current Equation

The solution of the current equation can be found in the same manner as the solution of the voltage equation with the exception of the initial conditions. We restate the current equation as follows:

$$\frac{d^2 I}{dx^2} = \gamma^2 I \quad \text{(A-15)}$$

The Laplace transform of the current as a function of distance is expressed as follows:

$$\mathcal{L}[I(x)] = I(s) \quad \text{(A-16)}$$

Taking the Laplace transform of the current equation yields

$$s^2 I(s) - s I(0+) - I'(0+) = \gamma^2 I(s) \quad \text{(A-17)}$$

The terms $I(0+)$ and $I'(0+)$ are initial conditions. They represent the constant terms which were present prior to the two successive differentiations of the original equation. The first initial condition $I(0-)$ represents the current at the source terminal of the transmission line which we will define as follows:

$$I(0+) = I_0 \quad \text{(A-18)}$$

The second initial condition $I'(0-)$ represents the rate of change of current with respect to distance along the transmission line at the source terminal. We will again assume no discontinuity of the transmission line at the source terminal so that $I(0+)$ becomes $I(0)$ and $I'(0+)$ becomes $I'(0)$. The rate of change of current with respect to distance along the transmission line was stated earlier to be

$$\frac{dI}{dx} = -VY \quad \text{(A-19)}$$

APPENDIX

The voltage at the source terminal of the transmission line was previously defined as

$$V(0) = V_0 \qquad (A\text{-}20)$$

The rate of change of current at the source terminal of the transmission line is therefore

$$\frac{dI}{dx} = -V_0 Y \qquad (A\text{-}21)$$

The initial conditions are then summarized as follows:

$$I(0) = I_0 \qquad (A\text{-}22)$$

$$I'(0) = -V_0 Y \qquad (A\text{-}23)$$

The Laplace transform of the current equation with the substitution of initial conditions can now be expressed as follows:

$$s^2 I(s) - s I_0 + V_0 Y = \gamma^2 I(s) \qquad (A\text{-}24)$$

Manipulating the equation to solve for $I(s)$ yields

$$I(s)(s^2 - \gamma^2) = s I_0 - V_0 Y \qquad (A\text{-}25)$$

$$I(s) = \frac{s I_0 - V_0 Y}{s^2 - \gamma^2} \qquad (A\text{-}26)$$

$$I(s) = I_0 \left(\frac{s}{s^2 - \gamma^2} \right) - \frac{V_0 Y}{\gamma} \left(\frac{\gamma}{s^2 - \gamma^2} \right) \qquad (A\text{-}27)$$

The solution of the current equation as a function of distance $I(x)$ may be found by the inverse Laplace transform of the function of s as follows:

$$\mathcal{L}^{-1}[I(s)] = I(x) = I_0 \cosh(\gamma x) - \frac{V_0 Y}{\gamma} \sinh(\gamma x) \qquad (A\text{-}28)$$

SOLUTION METHODS   221

A.2 METHOD OF SERIES EXPANSION
A.2.1 Solution of the Voltage Equation
The voltage equation is restated as follows:

$$\frac{d^2V}{dx^2} = \gamma^2 V \qquad \text{(A-29)}$$

The solution of the voltage equation may be expressed as the following series expansion:

$$V = a_0 + a_1 x + a_2 x^2 + a_3 x^3 + a_4 x^4 + a_5 x^5 + a_6 x^6 + \cdots \qquad \text{(A-30)}$$

The first derivative of the voltage may be expressed as the first derivative of the series expansion as follows:

$$\frac{dV}{dx} = a_1 + 2a_2 x + 3a_3 x^2 + 4a_4 x^3 + 5a_5 x^4 + 6a_6 x^5 + \cdots \qquad \text{(A-31)}$$

The second derivative of the voltage may be expressed as the second derivative of the series expansion as follows:

$$\frac{d^2V}{dx^2} = 2a_2 + 6a_3 x + 12a_4 x^2 + 20a_5 x^3 + 30a_6 x^4 + \cdots \qquad \text{(A-32)}$$

Since
$$\frac{d^2V}{dx^2} = \gamma^2 V \qquad \text{(A-33)}$$

Then
$$\frac{d^2V}{dx^2} = 2a_2 + 6a_3 x + 12a_4 x^2 + 20a_5 x^3 + 30a_6 x^4 + \cdots$$
$$= \gamma^2 (a_0 + a_1 x + a_2 x^2 + a_3 x^3 + a_4 x^4 + a_5 x^5 + a_6 x^6 + \cdots) \qquad \text{(A-34)}$$

The coefficients of the powers of x may be equated between the first line of the equation which represents the second derivative of the voltage with respect to distance and the second line of the equation which represents the product of the voltage and the propagation constant. The coefficients of x are summarized in Table A-1.

Table A-1 Corresponding Coefficients of x

POWER OF x	0	1	2	3	4	5	6, 7, $\cdots$
$\dfrac{d^2V}{dx^2}$	$2a_2$	$6a_3$	$12a_4$	$20a_5$	$30a_6$	$42a_7$	$\cdots$
V	$\gamma^2 a_0$	$\gamma^2 a_1$	$\gamma^2 a_2$	$\gamma^2 a_3$	$\gamma^2 a_4$	$\gamma^2 a_5$	$\cdots$

These coefficients can be further organized in terms of a_0 for the even powers of x and a_1 for the odd powers of x. Equating coefficients of the even powers of x yields

$$2a_2 = \gamma^2 a_0$$
$$a_2 = \frac{\gamma^2}{2} a_0 = \frac{\gamma^2}{2!} a_0 \tag{A-35}$$

$$12a_4 = \gamma^2 a_2$$
$$a_4 = \frac{\gamma^2}{12} a_2 = \left(\frac{\gamma^2}{12}\right)\left(\frac{\gamma^2}{2} a_0\right) = \frac{\gamma^4}{24} a_0 = \frac{\gamma^4}{4!} a_0 \tag{A-36}$$

$$30a_6 = \gamma^2 a_4$$
$$a_6 = \frac{\gamma^2}{30} a_4 = \left(\frac{\gamma^2}{30}\right)\left(\frac{\gamma^4}{24} a_0\right) = \frac{\gamma^6}{720} a_0 = \frac{\gamma^6}{6!} a_0 \tag{A-37}$$

Equating coefficients of the odd powers of x yields

$$6a_3 = \gamma^2 a_1$$
$$a_3 = \frac{\gamma^2}{6} a_1 = \frac{\gamma^2}{3!} a_1 \tag{A-38}$$

$$20a_5 = \gamma^2 a_3$$
$$a_5 = \frac{\gamma^2}{20} a_3 = \left(\frac{\gamma^2}{20}\right)\left(\frac{\gamma^2}{6} a_1\right) = \frac{\gamma^4}{120} a_1 = \frac{\gamma^4}{5!} a_1 \tag{A-39}$$

SOLUTION METHODS

Table A-2 Summary of Coefficients

POWER OF x	0	1	2	3	4	5	6, 7, $\cdots$
Coefficients of a_0	1	—	$-\dfrac{\gamma^2}{2!}$	—	$\dfrac{\gamma^4}{4!}$	—	$\cdots$
Coefficients of a_1	—	1	—	$-\dfrac{\gamma^3}{3!}$	—	$\dfrac{\gamma^5}{5!}$	$\cdots$

$$42a_7 = \gamma^2 a_5$$
$$a_7 = \frac{\gamma^2}{42} a_5 = \left(\frac{\gamma^2}{42}\right)\left(\frac{\gamma^4}{120} a_1\right) = \frac{\gamma^6}{5040} a_1 = \frac{\gamma^6}{7!} a_1 \qquad \text{(A-40)}$$

The even and odd coefficients are summarized in Table A-2 using factorial notation. The solution for V may therefore be written as

$$V = a_0 \left(1 + \frac{\gamma^2 x^2}{2!} + \frac{\gamma^4 x^4}{4!} + \frac{\gamma^6 x^6}{6!} + \cdots\right) + a_1 \left(x + \frac{\gamma^2 x^3}{3!} + \frac{\gamma^4 x^5}{5!} + \frac{\gamma^6 x^7}{7!} + \cdots\right) \qquad \text{(A-41)}$$

Combining terms,

$$V = a_0 \left(1 + \frac{(\gamma x)^2}{2!} + \frac{(\gamma x)^4}{4!} + \frac{(\gamma x)^6}{6!} + \cdots\right) + \frac{a_1}{\gamma}\left(\gamma x + \frac{(\gamma x)^3}{3!} + \frac{(\gamma x)^5}{5!} + \frac{(\gamma x)^7}{7!} + \cdots\right) \qquad \text{(A-42)}$$

The Maclauren series expansion of $\varepsilon^{\gamma x}$ is expressed as follows:

$$\varepsilon^{\gamma x} = 1 + \gamma x + \frac{(\gamma x)^2}{2!} + \frac{(\gamma x)^3}{3!} + \frac{(\gamma x)^4}{4!} + \frac{(\gamma x)^5}{5!} + \frac{(\gamma x)^6}{6!} + \cdots \qquad \text{(A-43)}$$

The Maclauren series expansion of $\varepsilon^{-\gamma x}$ is expressed as follows:

$$\varepsilon^{-\gamma x} = 1 - \gamma x + \frac{(\gamma x)^2}{2!} - \frac{(\gamma x)^3}{3!} + \frac{(\gamma x)^4}{4!} - \frac{(\gamma x)^5}{5!} + \frac{(\gamma x)^6}{6!} + \cdots \qquad \text{(A-44)}$$

APPENDIX

The sum of the series expansions $\varepsilon^{\gamma x}+\varepsilon^{-\gamma x}$ yields the even order terms of the series as follows:

$$\varepsilon^{\gamma x}+\varepsilon^{-\gamma x}=2\left(1+\frac{(\gamma x)^2}{2!}+\frac{(\gamma x)^4}{4!}+\frac{(\gamma x)^6}{6!}+\cdots\right) \quad (A\text{-}45)$$

The difference of the series expansions yields the odd order terms of the series as follows:

$$\varepsilon^{\gamma x}-\varepsilon^{-\gamma x}=2\left(\gamma x+\frac{(\gamma x)^3}{3!}+\frac{(\gamma x)^5}{5!}+\frac{(\gamma x)^7}{7!}+\cdots\right) \quad (A\text{-}46)$$

The solution for V may therefore be expressed as

$$V=\frac{a_0}{2}(\varepsilon^{\gamma x}+\varepsilon^{-\gamma x})+\frac{a_1}{2\gamma}(\varepsilon^{\gamma x}-\varepsilon^{-\gamma x}) \quad (A\text{-}47)$$

Since $\frac{1}{2}(\varepsilon^{\gamma x}+\varepsilon^{-\gamma x})=\cosh\gamma x$ and $\frac{1}{2}(\varepsilon^{\gamma x}-\varepsilon^{-\gamma x})=\sinh\gamma x$

Then
$$V=a_0\cosh(\gamma x)+\frac{a_1}{2\gamma}\sinh(\gamma x)$$

Evaluating initial conditions at $x=0$:

$$V(0)=V_0=a_0\cosh(0)+\frac{a_1}{\gamma}\sinh(0)=a_0(1)+\frac{a_1}{\gamma}(0)=a_0 \quad (A\text{-}48)$$

$$\frac{dV(0)}{dx}=-I_0Z=a_0\gamma\sinh(0)+\frac{a_1\gamma}{\gamma}\cosh(0)=a_0\gamma(0)+a_1(1)=a_1 \quad (A\text{-}49)$$

The solution of the voltage equation is therefore

$$V=V_0\cosh(\gamma x)-\frac{I_0Z}{\gamma}\sinh(\gamma x) \quad (A\text{-}50)$$

A.2.2 Solution of the Current Equation

The solution of the current equation by the method of series expansion is identical to the solution of the voltage equation with the exception of the initial conditions.

SOLUTION METHODS

$$\frac{d^2 I}{dx^2} = \gamma^2 I \qquad \text{(A-51)}$$

$$I(0) = I_0 = a_0 \cosh(0) + \frac{a_1}{\gamma} \sinh(0) = a_0(1) + \frac{a_1}{\gamma}(0) = a_0 \qquad \text{(A-52)}$$

$$\frac{dI(0)}{dx} = -V_0 Y = a_0 \gamma \sinh(0) + \frac{a_1 \gamma}{\gamma} \cosh(0) = a_0 \gamma(0) + a_1(1) = a_1 \qquad \text{(A-53)}$$

The solution of the current equation is therefore

$$I = I_0 \cosh(\gamma x) - \frac{V_0 Y}{\gamma} \sinh(\gamma x) \qquad \text{(A-54)}$$

INDEX

admittance, 65
angles
 current, 25, 26
 faulted bus currents, 26
 normal bus currents, 26
 phase, 25
 voltage, 25
asymmetrical current, 181
automatic generation control, 3

capacitance, 57
 energy in, 63, 65
 power in, 61, 62
 voltage and current relationships, 58, 59
capacitor bank, 110
characteristic impedance, 88
 lossless case, 91
 lossy case, 88
circuit elements, 45
conductance
 component of admittance, 66
 transmission line parameter, 67
conductivity, 46
cosinusoidal voltage and current, 19
current transformers,
 performance under fault conditions, 160
 performance under normal conditions, 160

demand, 114
digital fault recorder,
 frequency deviation, 6-9
 line-to-ground fault, 153
 line-to-line fault, 146
 real and reactive power measurement, 43
 third harmonic voltage, 32
 three-phase voltages and currents, 15

efficiency,
 of hydroelectric power plants, 37
 of transmission line operation, 112
electric service,
 demand component, 114
 energy component, 114
electrical length, 23, 24
electromagnetic radiation,
 velocity in free space, 23
 wavelength, 23
energy, 36, 42
 in capacitance, 63
 in inductance, 54
 in resistance, 175
Euler's theorem, 21

Fourier series, 30
frequency,

INDEX

bias mode, 3
deviation, 4-9
fundamental, 10, 30-32
radian 1-3
scheduled, 1, 3, 6, 9

harmonics, 7, 10, 30-32
and delta windings, 31
second harmonic, 10
third harmonic, 10, 30-32

impedance, 65
apparent, 44
characteristic, 88
discrete parameters, 66
lumped parameters, 66
negative sequence, 135
positive sequence, 134
zero sequence, 137
in-service testing, 44
inductance, 49
energy in, 54, 57
induced voltage, 49
power in, 53, 54
rate of change of flux
linkages, 49
voltage and current
relationships, 50, 51, 54, 55, 57

Laplace transform, 217-220
law of cosines, 27
load factor, 114
loads, 105
off-peak, 114
on-peak, 114

North American Electric
Reliability Council, 5
interconnection monitors, 5

regional monitors, 5
neutral currents, 10

off-peak, 114
on-peak, 114

power
and conjugate current, 39
and rms voltage and
current, 34
apparent, 35
complex, 39
hydroelectric, 36
instantaneous, 33-35
measurement of, 43
reactive, 35, 37
real, 35, 37
total, 38
transfer capacity, 38
units, 35
power circle diagrams, 117
derivation of equations, 118
power-angle diagram, 130
reactive power and voltage
magnitude, 125
real power and voltage
angle, 127
receiving bus phasors, 122
sending and receiving
diagrams , 123
sending bus phasors, 121
surge impedance loading, 128
transmission model, 117
power factor, 40
correction, 110
lagging, 41
leading, 40
propagation constant, 69, 91
distortionless case, 95

INDEX 229

lossless case, 94
lossy case, 92

radian, 2, 3, 24
reactance,
   capacitive, 60
   component of impedance, 66
   inductive, 51
reactor bank 176
resistance, 45-47
   component of impedance, 66
   critical, 166
   of aluminum conductors, 45
   of copper conductors, 46
   power in, 47, 48
   voltage and current relationships, 46
resistivity, 45
root-mean-square, 16, 20
   digital computation of, 16, 20

sampling rate, 7, 15
short circuits, 181
   constant flux linkages, 181
   derivation of the current equation, 182
   fault analysis, 187
   initial conditions, 181
   steady-state component of current, 181
   transient component of current, 181
sinusoidal voltage and current, 10
susceptance,
   component of admittance, 66
symmetrical components,
   applications to system protection, 159
   double-line-to-ground fault, 153
   fault point, 132
   line-to-ground fault, 146
   line-to-line fault, 139
   negative sequence components, 134
   phase quantities, 137
   positive sequence components, 133
   sequence quantities, 137
   viewed from the fault, 132
   zero sequence components, 136
system response,
   forced, 161
   natural, 161

time error,
   correction, 4, 5
   fast condition, 4, 5
   integrated, 4, 5
   negative, 4, 5
   positive, 4, 5
   slow condition, 4, 5
transformer inrush, 10
transformers,
   coefficient of coupling, 197
   coercive force, 206
   core design, 204
   core resistance, 200
   current relationships, 207-210
   current transformer model, 213
   delta-wye phase shift, 209, 211

INDEX

delta-wye transformations, 207-212
excitation, 204
flux density, 196, 197, 204, 205
flux relationships, 192-194
hystereses, 200, 205
ideal, 194
impedances, 135, 160
leakage flux, 192
magnetic flux density, 196, 197, 204, 205
magnetic flux intensity, 205
magnetizing reactance, 201
mutual flux, 193
polarity, 191
practical model, 200
primary and secondary quantities, 198
reflected quantities, 201-203
residual flux, 205
saturation, 206
self-inductance, 192
shell design, 204
transformer action, 195
turns ratio, 197
voltage relationships, 197, 210-212
winding reactance, 200
winding resistance, 200
transients,
critical resistance, 166
critically damped case, 166
damping ratio, 167
overdamped case, 172
undamped case, 163
undamped natural frequency, 163
underdamped case, 169
transmission lines, 67

ABCD constants, 84
approximate pi model, 87
approximate T model, 87
characteristic impedance, 88
charging, 97
development of the current equations, 70, 76
development of voltage equations, 68, 76
distance and time functions, 75
distributed model, 67
efficiency, 112
equivalent pi model, 84
equivalent T model, 84
general solution of equations, 71
initial conditions, 72
parameters, 67
particular solution of equations, 72
propagation constant, 69, 91
receiving-end voltage and current, 82
sending-end voltage and current, 83
series impedance, 67
series inductive reactance, 67
series resistance, 67
shunt admittance, 67
shunt conductance, 67
shunt susceptance, 67
solution of the current equations, 78
solution of the voltage equations, 78
telegrapher's equations, 78
terminal conditions, 81

traveling waves, 73
voltage regulation, 114

vector, 20
voltage and current phasors, 20

ABOUT THE AUTHOR

Geradino A. Pete is president of Vector Engineering Services, P.C., an engineering consulting firm in Clinton, New York. He is a member of the National Society of Professional Engineers and the Institute of Electrical and Electronic Engineers. Mr. Pete is also the author of *Substation Protection Philosophy*, *Voltage Control for Electric Power Systems*, and *Transmission Line Protection Philosophy*.